Time and Economics

Željko Rohatinski

Time and Economics

The Concept of Functional Time

palgrave
macmillan

Željko Rohatinski
Zagreb, Croatia

ISBN 978-3-319-61704-6 ISBN 978-3-319-61705-3 (eBook)
DOI 10.1007/978-3-319-61705-3

Library of Congress Control Number: 2017951018

Cover illustration: GoodGnom / Getty Images

Printed on acid-free paper

This Palgrave Macmillan imprint is published by Springer Nature
The registered company is Springer International Publishing AG
The registered company address is: Gewerbestrasse 11, 6330 Cham, Switzerland

To Leona

Foreword

From my first encounter with Željko Rohatinski, just over a decade ago, I was captivated by his unorthodox ideas that often ran counter to conventional wisdom. The meeting was in 2006 in the centuries-old seaside town of Rovinj on the Istrian coast. Rohatinski, then governor of the Croatian National Bank, was giving a lecture in which he warned of the dangers of bubbles in financial and real estate markets. Rohatinski had already been talking about the unsustainable financial and macroeconomic trends in the domestic and international environments for several years, emphasizing that central banks should focus on overall price and financial stability, but even in 2006 he remained one of the few economic leaders voicing this view. Two years later, the burst of these bubbles was a crucial factor in the 2008 global economic crisis.

Rohatinski's "out of the box" thinking has punctuated a long and illustrious career as an economist and policymaker in the public and private sectors. This was recognized when *The Banker* magazine, a publication of London's *Financial Times*, named Rohatinski Central Bank Governor of the Year of the World and Europe for 2008 for his pioneering work in the implementation of a series of unorthodox measures aimed at preserving financial stability. Stories circulating in the central bank at the time said he had a "magic drawer" in which he kept a set of targeted monetary and macroprudential policies, waiting for the right moment to use them.

One of the things hiding in that drawer was the preliminary version of this book. The idea had originated about 40 years ago when, as a doctoral candidate, Rohatinski contemplated the unusual idea of bringing Einstein's theory of relativity together with recognized models of economic activity. In many ways, this book symbolizes the personal journey of the author from a young economist inspired by theory to an experienced policymaker.

As editor of this book, I had the privilege of talking with Rohatinski about this journey and working with him to bring his unique concepts together into a concrete structure that could be shared with readers. Rohatinski acknowledged that he initially had no idea where his research would lead or of the potential implications of his findings. His contemplations were based primarily on the literature from several scientific disciplines, and the first phase was to explore how human understanding of time itself evolved from antiquity and Greek philosophy to modern quantum physics. In essence, he followed how various approaches to time affected the perception of different aspects of life through the ages. The next logical step was to review the perception of time and term determination of economic activity as seen by some of the great economists, such as Smith, Marshall and Keynes.

Rohatinski's deliberations culminated with the creation of a *functional* approach toward observing time, which confirmed that the dilation of time and contraction of lengths could be seen in the context of economic activity. In other words, by expressing all variables in Marx's two-sector model as a function of time, which is analogous to Keynes's model where variables were expressed as a function of wages, the author empirically proves time does not flow evenly in different systems. In the long run this affects not only the parameters, but also the variables of economic models. The main implication of these findings is that the same model cannot be used for different systems if they operate in different functional times. Rohatinski defines a very short period as a duration in which economic policy cannot have a significant impact on the efficiency of an economic system, and, at the opposite extreme, a very long period as one in which the overall organisation and the entire pattern of the economic activity changes, meaning that economic policy could shape both the qualitative and quantitative elements of an economic system.

Since Rohatinski developed this approach to the time dimension of economic activity, the world has changed immensely: new economic theories have emerged, data analytics have strengthened, and economic and financial systems across markets have become increasingly integrated. At the same time, the author himself has been occupied solving real economic problems as a macroeconomist and central banker. The experience has given him a deep practical understanding of how economic systems function, which has leavened and enriched his initial theoretical deliberations. After such a long and turbulent period it is fair to ask whether time itself has overtaken the young doctoral candidate's ideas of the time dimension of economic activity. But, augmented by intense professional experience, Rohatinski's main concepts about the importance of the time dimension in better understanding economic activity are even more important as they provide policymakers with useful insights on how to influence that activity and directly affect the potential efficiencies of micro- and macroeconomic policies and subjects in different time periods.

Rohatinski's *Time and Economics* joins ideas from philosophy, physics and economics into an original concept, challenging accepted presumptions of contemporary economic science. Amid today's dynamic flows of goods, labour and capital around the globe and the rapid advancement of technology, this book provides an essential foundation for policymakers and economists working to guide the world to a more prosperous and equitable future.

Mirna Dumičić

Preface

As a young economist more than 35 years ago, I was convinced that a broad area in economic science remained insufficiently explored. Not only in the context of its application, but also within its fundamental ideas and their interaction with the findings of other sciences. Following through on these impressions, I enthusiastically studied the very foundations of economics presented in the classical political economics framework and above all of in its quintessential work, Karl Marx's *Capital.* I was not encumbered by the ideological hypotheses and consequences resulting from Marx's labour theory of value, but instead focused on the construction of his model of capital reproduction. Even though this was a theoretical model without strong empirical foundation, it clearly pointed to the essence of the causal, structural, and equilibrium (and non-equilibrium) relations among the economic categories in real economic life.

The problem that I found particularly intriguing was the meaning of time in which these relations are established and whether any of the theories that explain time as a relative category, rather than as an absolute category, could be applied to these relations. This trail led to a complex research question: *Would it be possible to identify an "economic time" as a factor that directly affected the modalities in which the economic activity of individuals and society took place?* My initial attempt to explore this issue resulted in a doctoral dissertation, "Role, Goals and Methods of Economic

Policy and the Maturity Problem," which I forgot soon after it was accepted in 1988.[1]

In the meantime, I worked as a macroeconomist in public administration, in an economic institute, and in commercial banking, as well as serving as central bank governor for Croatia from 2000 to 2012. These endeavours occupied me with solving "everyday" economic problems and took me far away from the topics and research preferences of my youth.

Economic science over those years changed immensely, in terms of methodology and analytics, as well as theory. Supply-side economics, new neo-classical synthesis, theory of rational expectations, theory of real business cycles, effective markets hypothesis, and modern monetary theory are just some of the prevailing theories that have had an essential impact on the formation of contemporary economic science, while the development of computer and information technology has enabled their hypotheses to be verified and increasingly sophisticated empirical models to be constructed.

In parallel, a period of Great Moderation, an easing of economic volatility, encouraged the broad impression that economic science had found credible answers to the majority of the crucial economic issues of modern society. However, the global financial crisis of 2008 and its aftermath, including secular stagnation, proved this broad impression was illusive. This was not only because the complexity of economic processes and their non-economic determinants are difficult to model empirically, but also because of the simplifications of some of the ideas and premises on which these models were founded.

Economic science attempted to respond to some of these concerns by re-examining and re-designing existing models, constructing new models, or creating some hybrid of old and new, as well as by restricting the applicability of some models to specific cases. But the problem that calls the foundations of economic models into question is much more complex and goes back into the essence of "economic philosophy" and the qualitative and quantitative categorical apparatus that was built up during its historical development. Re-examining this apparatus, parts of which are often perceived as axioms, would mean re-thinking economic science from the start and determining the influence of findings from other areas of science on its formation. This might be seen as superfluous if it did not have an applicative value in practice.

After re-reading my dissertation with a greater awareness of the theoretical, empirical, and real-life problems existing in contemporary economic science, it seemed that many of the arguments and ideas of classical political economy still may be quite relevant. In addition, the re-interpretation of its initial premises and key postulates of the elements of political economy could inspire future economic research. This encouraged me to try to share these deliberations with a wider public.

The main purpose of this book is to introduce, explain, and demonstrate the importance of determining the time dimension of economic activity. This book employs an understanding of time as a relative concept against a background from modern physics and philosophy, and it shows that such an approach applied to economics enables better comprehension of the forms and modalities of economic activity.

Definitions and understanding of time through the history have been presented through works of Aristotle, Descartes, Newton, Boscovich, Einstein, Bloch, and others who observe it as an objective category, as well as through the conclusions of Plotinus, Augustine, Leibnitz, Kant, Heidegger, Husserl, Merleau-Ponty, and others who perceive it as a subjective category. In questioning the problem of absolute (Newton) and relative time (Einstein), special focus is put on Whitehead's philosophical description of time relativity, which analysed the space-time structure of an individual event, as well as the impact of that structure on relations of this event with other events.

Since none of these authors explored the problem of time in the context of economic activity, this book augments this part of the literature as it elaborates on the application of relative time in economics. Time is observed as an objective functional category that is the integral element of economic activity and one of its most important modulators.

The starting point for the constructed theoretical model has been the classical political economy framework presented in Marx's *Capital*, to which the concept of relative time defined in modern physics and philosophy was applied. For the purpose of exploring time dimension of economic activity, Marx's model is mathematically described and expanded by the variable "time," not as an absolute category, but as a relative or functional one, meaning that all variables are expressed as a function of time (which is analogous to Keynes' model, in which variables were expressed as function of wages).

In this sense, this book links Einstein's special theory of relativity with economic processes. The economical phenomena observed in this model—*the appearance of dilation of time and the contraction of costs in economic systems and sub-systems that move relative to each other*—are analogous to the ones in physics—*dilation of time and contraction of lengths of subjects in the inertial reference frames that move relative to each other* that are described by modified Lorentz transformations (such as Einstein, 1950, and Ugarov, 1979). This implies different capital efficiency in these two systems, despite the equal initial conditions of their reproduction. Such an understanding of time has direct implications on modalities of economic activity and should be taken into account when considering the time dimension or maturity of economic activity, for instance in creating economic policy.

The literature specifically devoted to the time dimension of economic activity is relatively scarce, but there are several researchers that explore some segments of the problem presented here. In 1950, English economist George L. S. Shackle was one of the first economists to argue that economics that use equilibrium methods ignore the dimension of time and who questioned the validity of the mechanical time dynamics models. He focused on the dynamic movement in time: translation of the moment-in-being along the calendar axis (outside) and from one moment in to another (inside). A recently published analysis of Shackle's work (Madsen, 2015) additionally pointed to the specific nature of time in economics compared to other sciences.

In 1973, Branko Horvat, one of the most prominent Croatian economists, explored the problems of contraction of costs of fixed capital and dilation of "economic time" in evenly growing economic systems and showed that every economic system has its own, inherent "economic time," which is in line with the conclusion reached in this book. By contrasting the atemporal Marshallian model with an explicit time model with uncertainty about costs at the firm level, the American Nobel Laureate Peter A. Diamond in 1994 focused on modalities of establishing micro- and macroeconomic equilibria in periods of different durations.

Later in 2010, another American Laureate, Paul Krugman, examined relative time. Using interest rates on goods in transit travelling at close to the speed of light as an example, Krugman explored the problem of dilation of time and contraction of lengths in a system that moves relatively with respect to another system.

While these studies explored some specific aspects of time in economics, this book provides a comprehensive approach towards analysing the relationship between time and economic activity.

This research also examines the definitions and classifications of time periods used by Smith, Marshall, and Keynes. Although these approaches differ significantly, they all point to the essential problems related to the attempts to "fit" economic activity in the standard units of calendar time during the process of establishing partial or general equilibrium in the economic system. This problem is even more complex in the context of all causal, structural, and (dis)equilibrium relations between economic categories. Based on this review, this book offers a concept of looking at time periods in that context—the "functional approach."

Contents

About the Author

Former Croatian National Bank Governor Dr. Željko Rohatinski is a Croatian macroeconomist who has worked in government administration, economic institutes, commercial banking, and the corporate sector. Among numerous recognitions, he was named The Best Central Bank Governor of the World and Europe in 2008 (The Banker magazine), The Best Communicator of the Year 2008 (Croatian Public Relations Association), and the Most Respectable Financial Expert (Lex Info Award).

Mirna Dumičić is an economist who currently works as a financial stability expert in the Macroprudential Policy Division at the European Central Bank, where she came in October 2016 from the Financial Stability Department of the Croatian National Bank.

Roger Malone is a former bureau chief for Vienna and Singapore for Dow Jones Newswires and former senior editor for McKinsey & Co. He currently consults on editorial strategy and execution from his base in Zagreb, Croatia. He holds a master's in International Relations and Economics from The Johns Hopkins University School of Advanced International Studies.

Notes

1. The integral version of my doctoral thesis, entitled "Vremenska dimenzija ekonomske aktivnosti drustva," was published in Croatia in 1990 by the Institute of Public Finance.

List of Figures

List of Tables

1

Introduction

> *One reason why modern life is so uncomfortable is that we have grown self-conscious about things that used to be taken for granted.*
>
> Joan Robinson, *Economic Philosophy*[1]

From its classical origins, economic theory has been frozen in calendar time. The discipline's ideas of how people, markets, and nations produce, distribute, and consume have evolved, but critical aspects of time itself have remained underexplored.

The concept of time in economics has not kept pace with that of other sciences, and economists continue to see their world through the lens of calendar time, which blurs the perception of economic activity and distorts the parameters of economic models. By looking at time using the kaleidoscope of ideas developed by writers in philosophy and physics and joining these concepts to economic activity, economists can create an understanding of functional time, which could lead to more effective and beneficial public policies.

From physics, the theory of relativity says people travelling at different speeds experience time differently. In much the same way, different eco-

Ž. Rohatinski, *Time and Economics*, DOI 10.1007/978-3-319-61705-3_1

nomic players or entities experience functional time differently based on the duration of their unique economic cycles or their own speed of production. A better understanding of the relationship between time and speed in micro- and macroeconomic production—drawing from modern theories in economics, physics, and philosophy—can help avoid distortions in public policy that contribute to inefficiencies and waste.

Economic science analyses the flow and modalities of economic activity from a short- and long-term perspective. In the short-term view, the processes of establishing supply–demand equilibrium on partial markets of products, capital, and labour, as well as the modalities of establishing general systemic equilibrium, are considered. The long-term approach is primarily focused on economic growth, allocation of the factors of production, and meeting key economic preferences of a society. Such division of economic analysis is based on the classification of time periods of different durations, depending on the modalities of the flows of reproduction in a specific time period. Naturally, this classification depends on understanding of the nature and character of time and, by extension, the manner in which economic activity takes place in time.

However, today's economic understanding of the nature of time is largely based on the views of classical physics about the absolute character of time, space, and motion, while most modern physical and philosophical theories define time as a functional and a relative category.

From an economics point of view, this raises several crucial questions:

- Is it possible to apply an understanding of time formulated within the framework of modern physics and philosophy when considering the time dimension of economic activity?
- Does such an approach enable a better understanding of the forms and modalities of economic activity?
- Could such an understanding of time directly affect how the modality in which economic activity takes place is established and changes, and if so how?
- How is this time manifested under this perspective, and how is it measured?
- Is it possible, in line with the special theory of relativity and Lorentz transformations, to identify within economic theory phenomena

analogous to the dilation of time and the contraction of lengths in physics?

- How would this understanding of time or the time flow of economic activity be reflected in considering the temporal determination—"maturity" or "duration"—of this activity?

A search for answers to these questions requires a consideration of the basic postulates of the physical and philosophical understanding of nature and the character of time, as well as the contents, characteristics, and modalities in which economic activity occurs. The elements essential for determining the time dimension of this activity can be derived only within the intersection of these two views.

Part I: The Functional Time Hypothesis

The first part of this book reviews the development and basic postulates of selected physical and philosophical theories about the nature and character of time and its perception. These theories represent the foundation for setting the initial hypotheses from which the features and influence of the time component of economic activity will be elaborated.

Following that, two groups of theories that differ primarily according to the character they assign to the category of time are presented. The first treats time as an objective category defined by natural laws of motion and evolution of substance, independent of the perception of the observer, and is represented through the works of Aristotle, René Descartes, Isaac Newton, Roger Joseph Boscovich, Albert Einstein, Alfred North Whitehead, and others (Chap. 2—Time as an Objective Category). The second observes time as a "horizon of perception" and a cognition of being and references conclusions reached by Plotinus, Augustine, Gottfried Leibnitz, Immanuel Kant, Martin Heidegger, Edmund Husserl, Maurice Merleau-Ponty, and others (Chap. 3—Time as a Subjective Category).

Within this framework, the question of the absolute (Newton) or relative (Einstein, Whitehead) character of time is discussed in more detail. The essence of this dilemma is whether the time is unaffected by events and processes that occur within it (Is it universal and does it flow evenly?) or do these events and processes have an essential effect on the formation and course of time, which then become distinctive (Do events affect the nature of time?).

Notwithstanding the essential differences between the theories presented, Part I ends with a synthesis of elements that profile an understanding of time as a functional category. These elements are presented to allow the consideration of the time dimension of economic activity, taking into account the specific features of that activity in the context of natural and, in particular, physical events and processes. Part I concludes that a subjective approach to time provides a static awareness of sequence and duration, but this becomes dynamic once an observer in objective time enters the equation.

Part II: The Structure and Character of Economic Activity

The second part of the book elaborates on the issues of the nature, structure, and characteristics of economic activity, which together determine the time dimension of that activity. It begins by elaborating on two issues: the functional structure and the circular flow of economic activity. To determine the basic elements and ways in which they are functionally incorporated into economic activity, the discussion on the functional structure considers the content and mutual relations of economic events, processes, and the economic activities of individual subjects, while the discussion of the circular flow considers a periodicity of that activity or, in other words, consecutive repetition of its basic entities as means and form of the reproduction of economic categories and economic subjects.

First, Part II reviews the historical context of economic activity (Chap. 4—Historical Character). Despite numerous definitions of "economic activity," which vary widely based on the specific scientific discipline, religion, or ideology of the writer, there is a broad agreement on the essential influence economic activity has had on the development of the society and relations between its subjects. Following the continuous organisational and structural changes of a society, a historical character of economic activity is implied, and this character has also been evolving and changing in terms of content, forms, and characteristics because of its interaction with the material and spiritual evolution of society's production capacity, as well as its ethical, political, and social norms.

Within the framework of its historical development, economic activity is circular and comprised of production, distribution, exchange, and consumption (Chap. 5—Circular Character). These processes are continuously reproducing in an equivalent (simple reproduction) or changed capacity (extended reproduction or recession).

A contemporary context is also presented (Chap. 6—Components and Course). Here terms and concepts essential for understanding the role and character of "economic time" are defined:

- Event or process for which the time dimension is analysed.
- The rule under which this event or process unfolds and which defines its periodicity or its internal time measure.
- Human perception of these events or processes and rules, which define the observation of their sequence and length.

This chapter also explains the main types of functional relationships established between economic events or processes:

- *Causal relations* between different economic events, when one event is caused by a previous event and reflected in subsequent events.
- *Vertical relations* that reflect the interaction of economic events that are part of the same area of economic activity, but include participants from different hierarchical levels.
- *Horizontal relations* that reflect the aggregation of analogous economic events that could be observed from an individual point of view or at the level of the whole society (aggregated production or consumption of economic subjects).

The analysis shows that economic activity is complex and occurs on multiple levels, allowing individuals and societies to organise it in a way to promote their existence and development. Economic activity itself comprises events, which are relationships among two or more economic subjects, and processes, which are functional rounded groups of events. The confluence of these processes creates the circular flow of economic reproduction.

Part II continues by addressing the circular flow of reproduction of economic categories and the modalities of the functioning of the economic system. These modalities are observed through a form, type, and manner in which a complex set of qualitative and quantitative relationships among economic categories in the circular flow of their reproduction are established.

Society has numerous natural, technical, financial, and other resources that form the basis for pursuing economic activity. When included into economic events and processes, these resources become the corresponding economic categories. Consequently, a complex group of economic relationships is created among these categories:

- *Causal relationships*: An essential characteristic of any economic category, which determines the main mechanism for establishing functional relations between the economic events and processes by way of qualitative-quantitative transformations of economic categories into a unique course of the economic activity of the certain subjects and a society (Chap. 7—Causal Relations; Appendix A—Transformation Problem).
- *Structural relationships*: Established between the functionally related economic categories during an economic event or in the same phase of a circular flow of economic activity. These relationships also represent a basis of homogenisation or diversification of individual circular flows of reproduction of economic categories (Chap. 8—Structural Relations).
- *Equilibrium relationships*: A regulated set of causal and structural connections between economic categories, ensuring their regular reproduction and economic activity (Chap. 9—Equilibrium Relations).

This re-examination of the usual definitions and exploration of relations among economic categories is followed by an analytical consideration of the properties that these relations should have for the normal process of reproduction of economic categories (Chap. 10—Modelling Economic

Activity). The structural model presented for reproduction in a two-sector economy is modified from Karl Marx's diagrams of simple and expanded reproduction. This modification centres on the selection of the functional unit of observation of the course of reproduction in a given sector and in the system as a whole. While Marx uses turnover of the capital advanced to the first and second sector, assuming an equal duration for each flow of reproduction, in the model presented here the unit of observation is a single circular flow of reproduction of this capital. This enables a better analytical comprehension of the course of the reproduction of economic categories and of the economic system itself.

Finally, Part II considers the character and behaviour of general rules of the functioning of the economic system and the reproduction of economic categories, as well as the information system through which economic subjects direct their activities towards the realisation of their preferences and motives for economic activity. This approach identifies and describes all the elements of the economic activity (fundamental entities, the periodicity in which they unfold, the modalities of the normal functioning of an economic system, and the information base) needed for a consideration of the temporal dimension of that activity.

Based on this, a model of simple and extended reproduction in a two-sector economy is constructed and conditions for its equilibrium defined (Appendix B—Conditions for Moderate Growth in a Two-Sector Economy; Appendix C—Conditions for Normal Reproduction in a Two-Sector Economy).

Part III—Economic Activity in Time and the Character of Functional Time

Part III merges the views on nature and character of time with those of the essential elements and properties of economic activity and analyses the time dimension of economic activity of individual subjects and of a society. Special consideration is given to two basic aspects: the nature, character, and impact of "economic time" on the modalities of economic activity (the reproduction of economic categories and economic systems) and the time definition of this activity as dependent on the modality of its unfolding in periods of economic time.

The section begins with an analysis of the temporal flow and temporal attributes of an individual or sub-system in the composition of a more complex economic system and defines the concept of the functional time of reproduction of economic categories in that system (Chap. 11—Functional Time). The model of reproduction in a two-sector economy is used to analyse the relationships of functional temporal flows of reproduction of economic categories in these sub-systems and the impact of the constellation of these temporal flows on the formation of causal and structural relations among the economic categories in the whole system while in equilibrium. Within this framework, a separate analysis is made of the length of the time period, expressed in units of functional time of the sub-systems of the economic system required for achieving the system equilibrium.

Four themes are highlighted from this analysis. First, while a single circular economic flow is defined by calendar time, processes within that cycle vary substantially in the duration of working periods, producing units that can be used to express the duration of the cycle. Second, because durations of working periods vary, turnover and functional time can differ when measured by this unit, creating varying densities of turnover time. Third, to reconcile these differences and bring capital reproduction into a synchronous cycle, specific constellations of causal and structural relationships must be established. And finally, each economic system with a macroeconomic whole proceeds at its own functional time, which influences its own features and the overall macroeconomic system as well (Appendix D—Conditions for Equilibrium of a Two-Sector Economy in Functional Time).

However, these activities must be reconciled with calendar time, which obscured relevant insights into the nature of functional time and created a distorted image of the efficiencies of economic activities. Without an understanding of the different densities of events and processes within calendar time, a comprehensive view of the functional flow of economic time is impossible. By aggregating varying durations into rigid calendar time terms, the true parameters of economic variables are distorted. And, finally, because calendar time units transect individual functional units differently, perceptions of functional relationships are also twisted.

Part III also examines the problem of different efficiencies among sub-systems and the entire system in terms of units of functional time (Chap. 12—The Relative Character of Functional Time), looking specifically at relativity of economic time and the apparent dilation of time and contraction of costs among economic systems and sub-systems that move relative to each other. This phenomenon implies that capital efficiency in one system is different when compared to that of another. This is analogous to Einstein's explanation of the dilation of time and contraction of length of subjects in the inertial reference frames that move relative to each other and can be described by modified Lorentz transformations. In this sense, when there is no variable constant across in all reference frames, such as the speed of light in physics, the determinants of the Lorentz transformation factor are variable both in space and time.

In its final chapter, Part III critically reviews the procedures and criteria for the classification of time periods of various lengths depending on the characteristics and modality of the economic activity that comprise each, referencing the works of Adam Smith, David Ricardo, Alfred Marshall, and John Maynard Keynes and also in the context of modern micro- and macroeconomic analysis and theory (Chap. 13—The Time Determination of Economic Activity).

Analysing the elements that constrain the operative application of these procedures and criteria shows the issue of the time definition of economic activity should be considered from the aspect of complex quantitative and qualitative features of that activity and changes to these features during this activity. Therefore, Part III concludes with a functional classification of periods of time and proposes criteria for the determination of the temporal definition of economic activity that contain a functional understanding of the nature and character of economic time and take into account the complexity of the content and modality of the unfolding of the economic activity in this time.

The main presentation ends with concluding remarks.

Notes

1. Joan Robinson, *Economic Philosophy* (Bungay, Suffolk: Penguin Books, 1962) 1.

Reference

Robinson, Joan. 1962. *Economic Philosophy*. Bungay, Suffolk: Penguin Books.

Part I

The Functional Time Hypothesis

Human recognition of time predates history. Without words like *hour* or *month* to contain the concept, women and men followed the passing of time in the migratory patterns of their prey, the annual ripening of the flora around them, and the cycle of child birth, among countless other observations. At the dawn of written history, under the Greek *chronos* and the Latin *tempus*, mankind began categorising time as an element typical for all natural and social processes, with its first explicit definition emerging from Greek philosophical thought.

In his dialogue *Timaeus*, written in the fourth century BC, Plato defined time as an integral element of the harmonious structure and organisation of the world, representing one of the principal distinctions between chaos and cosmos. Time originated with the universe and exists together with it. Further, time represented a stimulus of permanent and invariant eternity and contained the dimensions of the past, present, and future. Through its cyclical flow, dependent on the motions of celestial bodies, time gravitated towards a numerical measure representing a chronological or chronometric category.

In the millennia following this early attempt to capture the essence of time, two predominant approaches to understanding time have emerged: the objectivist and the subjectivist. According to an objectivist approach, time was an objective category defined by natural laws and was independent of the subjective perception of the observer. The subjectivist approach implied that this objective or mechanical conception of time was merely an aspect of incidence. The essence of time, a subjectivist would argue, was found in internal or real time, which encompassed a horizon of perception and the cognition of being.

In practice, the distinction between these two approaches to understanding time has not always been precise or constant, and indeed in some cases the elements of both approaches have been conflated into one synthesised view. This trend was especially true in the late seventeenth and early eighteenth centuries as authors joined together elements of physics, philosophy, and theology to begin the journey towards what we see as functional time.

2

Time as an Objective Category

In its original form, the objectivist approach to time was derived from Aristotle's dynamic definition of time as a measure of motion towards the past or towards the future. Here, time was understood as a quantitative category expressed by the difference between the two states and the motion of matter. Within this definition, however, time was neither motion nor change, since these aspects manifested themselves in something that moves and changes, while time flowed evenly at all points. Aristotle (1952) emphasised that time did not exist without change and motion: it existed *in* motion, but was not motion.

Move ahead 2000 years and, in the works of René Descartes and Isaac Newton, the concept of objective time has been accepted as the key principle of structure and existence of the universe. Descartes saw time as a category originating together with the universe, which was manifested locally through the motion of celestial bodies and expressed the duration and sequence of the occurrences, while, for Newton (1962), time represented a category of its own, unaffected by the existence of the universe and the processes within it. In a word, it was absolute:

> *Absolute, true, and mathematical time, of itself, and from its own nature, flows equably without relation to anything external, and by another name is called*

Ž. Rohatinski, *Time and Economics*, DOI 10.1007/978-3-319-61705-3_2

> *duration: relative, apparent and common time, is some sensible and external (whether accurate or unequable) measure of duration by the means of motion, which is commonly used instead of true time: such as an hour, a day, a month, a year.*[1]

These measures are by necessity imperfect since, as Newton wrote, "it may be that there is no such thing as an equable motion, whereby time may be accurately measured. All motions may be accelerated and retarded, but the flowing of absolute time is not liable to any change."[2]

By integrating the concept of absolute time with that of absolute space, all within a framework of absolute motion, Newton established the theory of an absolute space-time universe—eternal and infinite—the wasn't affected by the existence, development, or motion of celestial bodies and human cognition.[3] Within this universe, a unique unaffected time exists for all reference frames. In these conditions, the body, occurrence, or process transferring from one spatial system to another, which is specified by the three-dimensionality of space, retained a unique time feature and a unique time dimension of existence encompassed by a concept known as Galilean transformations. On this basis, Albert Einstein concluded that Newtonian mechanics were relative in relation to the reference frame. Therefore, two non-simultaneous events occurring in the same place have no objective meaning and are independent from the reference frame.

The absolute character of space and time, which Newton used as a basis for his laws of motion, has thus become the one of the main postulates of the classical physics and the prevailing mode of thinking about the essence and character of the time category. This approach has been dominant through to Einstein's special theory of relativity, even as some respected authors questioned its validity, both partially and as a whole.

Among the first to raise questions was the eighteenth-century Croatian physicist and philosopher Roger Joseph Boscovich. Although German physicist Werner Heisenberg, who described Boscovich as "the Croatian Leibnitz," said his teachings on the relativity of space and time and how they change in relation to motion were not expressed with sufficient clarity, in some respects Boscovich can be seen as a predecessor to Einstein.

Without completely denying the absolute character of time, space, and motion, Boscovich in his 1758 volume, *Theoria Philosophiae Naturalis*, and especially in its subsequent amendments, *Space and Time as They Are Perceived by Us*, conjectured that it was necessary to admit that "there is some real mode of existence" that demands that a "thing is where it is and exists then, when it does exist." These aspects were the real time and real place that were individually created and individually destroyed and were completely indivisible, non-extended, motionless, and invariant in their order, while every point of matter, provided it exists, necessarily related a point in space with a moment in time.[4]

The point in space and the moment in time were real and provided the basis for an actual relation of distance—physical or local distance between two points or temporal distance between two events. That there was a specific distance between the two points essentially meant that the given points have their specific modes of existence, which were necessarily changed with the change of this specific distance. Considering the local structure, Boscovich referred to these modes as real local points, and with regard to the temporal structure he referred to them as moments; real local points lacked any extensibility, while moments lacked any duration. Both were deprived of any divisibility.

Boscovich believed between these points it was always possible to insert another midway, which will touch neither of the former. On this basis he concluded that although the number of interpolated or extrapolated points of matter was finite, the distance between them was temporally and spatially infinite. Therefore, any point of matter may occupy an infinite number of locations and temporal moments and have an infinite number of modes of existence. Its spatial and temporal determinants specified its imaginary time and space, which were motionless, infinite, and perpetual. Spaces and times belonging to all points of matter collided and represented one space and time.

As presented, the basis for Boscovich's viewpoint diverged significantly from that of0 Newton's theory of absolute time and space, which was unchangeable and unaffected. Boscovich's viewpoint referred to changes in time and space that resulted from the motion of matter, which he illustrated with an example of ten-foot rods used as measuring units. Boscovich began by asserting it is always possible to insert a third point between any two points of matter, implying that infinite distance existed between any two points and therefore matter was not continuous, but discrete. Boscovich linked this to the idea that transferring the points of matter changed the mode of their existence and thus changed their absolute distance. He concluded that transferring the points of matter from one position to another also changed the mode of existence of the measuring unit. This measuring unit could no longer present the uniformity of distance between the points of matter in one position in relation to another position in space, and, analogously, in time.

Further, Boscovich pointed out:

> *Just as with the measurement of time…, with the measurement of space it is impossible to transfer a fixed length from its place to some other, just as it is impossible to transfer a fixed interval of time, with regard to so that it can be used for the purpose of comparing two of them by means of a third. In space, both cases, a second length, or a second duration is substituted, which is supposed to be equal to the first; that is to say, fresh real positions of the points of the same ten-foot rod which constitute a new distance, such as a new circuit made by the same rod, or a fresh temporal distance between two beginnings and two ends.*[5]

Here Boscovich described precisely the idea of a "certain" relativity of time, space, and motion—later elaborated upon and mathematically shaped by the Dutch physicist Hendrik A. Lorentz—and provided one of the foundations for the modern theory of relativity. Boscovich's contribution to the theory of relativity is primarily reflected in the formation of an idea, but because of his place in the timeline of scientific thinking he could not elaborate upon it sufficiently to question Newton's conception of absolute time, space, and motion. This is why Boscovich treated the relativity of time, space, and motion, as well as the existence of multiple times and spaces, as "imaginary" categories: they were potential real time and space, accepting Newton's view on absolute simultaneity.

For more than 150 years, Boscovich's understanding of time and space remained on the fringes of accepted scientific thinking and left out of contemporary considerations of the subject. Throughout the nineteenth century, mainstream scientific thought was dominated by the conflict between the idealist and materialist perspective of the universe. Reflecting debates on the categories of time and space, this conflict centred on the disparity between understanding these categories as subjective and objective. Representatives of the idealist view, especially the Austrian physicist Ernst Mach, treated time and space as relative, rather than absolute, categories, and denied their objective existence. On the other hand, the classics of dialectical materialism rejected the idealist view of time and space as subjective categories, returning to Newton's conception of absolute time and space free of any theistic views.

This debate highlighted the disparities in understanding of objective or subjective character of time and space and also raised a series of other questions. Among these, special importance was given to a specific query: If time were an objective category, was it necessarily an absolute category as well, or could understanding time as a relative category also align with its objective character?

Even Einstein fell short of a uniform and decisive answer to this question. In his special theory of relativity, partly influenced by Mach, Einstein established a physical conception of the relativity of time. In this context, the theory itself comprised no elements that would explicitly question the objective character of time. Einstein also emphasised in his later works that the spatial and temporal data do not have fictitious, but real physical meaning. However, the essence of his understanding of the category of time, which was based on the relative simultaneity of events, suggested a multitude of space-time systems in which the objectivity of time and the subjectivity of the observer's perception could not be fully distinguished.

The initial basis of Einstein's understanding of the relativity of time, space, and motion was represented by the two fundamental principles of his special theory of relativity:

- All identical physical phenomena occurred equally in inertial reference frames provided equal initial conditions, which meant that there is no "privileged" reference frame.
- The speed of light in a vacuum was equal in all directions and in any part of the given reference frame and was equal in all reference frames, while the speed of signal transfer was finite, not infinite as provided by Newton's theory.

This was a direct basis for Lorentz transformations of spatial and temporal coordinates as a fundamental presentation of the concept of relativity of time, space, and motion.

Unlike Galilean transformations, Lorentz transformations implied that for two inertial frames that move at a constant velocity[6] relative to each other along one of the space coordinates and in which there are identical and synchronised measures of length and time, two simultaneous events in one reference frame would not be observed in the same way from the second reference frame. This suggested that the motion of the frame impacted the flow of time and that time flows in the reference frames moving relative to each other were different. When applied to relations between two non-simultaneous events and moving objects, Lorentz transformations were used to directly derive the two key conclusions of the theory of relativity of time, space, and motion: the phenomena of time dilation and, analogously, length contraction. In this understanding, time flowed more slowly in a reference frame moving relative to the reference frame of the observer, along with the "simultaneous" shrinking of lengths in direction of movement of an object.

This challenged the two main hypotheses of Galilean transformations:

- Absolute time, in which an instant of a certain event in one reference frame was identical to an instant in another frame.
- Absolute distance or length, in which the distance between the two points or two events was identical in both reference frames moving relative to each other.

Apart from the event itself, neither a point in space nor a moment in time in which something occurs possessed physical reality. As a result, there was no absolute relation in space that would be independent of the reference frames and no absolute relation in time between the two events. Based on this, Boscovich concluded there was instead only one relation in space and time independent of the reference frame.

Apart from offering a different definition of time flow to Newton's conception, this theory described time flow with essentially different content and character. According to the theory of relativity, the temporal "coordinates" of a certain event embodied no characteristic or, as defined by Boscovich, no mode of existence independent of the space coordinates, but together these coordinates formed a uniform space-time continuum. On this basis, German mathematician Hermann Minkowski replaced Euclid's three-dimensional space with a four-dimensional system of space-time, with the fourth dimension representing a product of time and the speed of light (Fig. 2.1).[7]

The theory of relativity distinguished two types of time or two modes of manifestation and flow of time:

- Object's time: time in a system moves together with an object.
- Coordinate time: time dilation occurs in a system moving relative to the first system.

This theory has had significant influence across a wide range of modern natural sciences, including philosophy, with scientists attempting to work out its main principles and apply them in analysing concrete relations in nature and society. Understandings reached in this manner lost

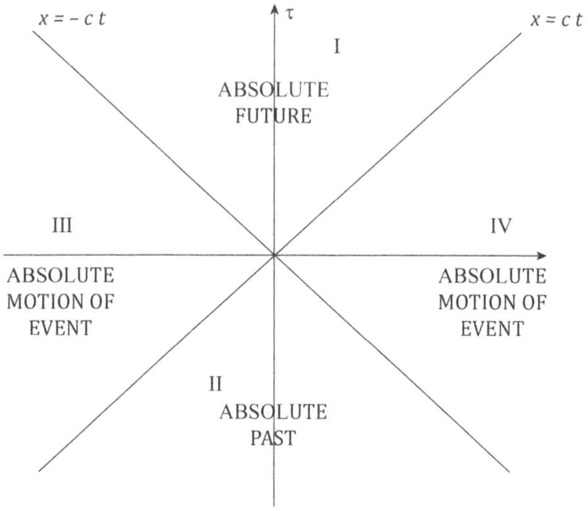

Fig. 2.1 Minkowski's time-space diagram. Source: Ugarov, *Special Theory of Relativity*

their direct connection with the usual empirical analysis of specific processes, including the hypothesis and principles of classic science based on such empirical approach.

Within this framework, the common understanding of time and space was abandoned, as was distinguishing between time as an exclusively objective or an exclusively subjective category. Further, the essence and genesis of time were no longer sought in the external environment of natural or social occurrences—such as in the absolute character of the matter, spirit, or observer perception—but rather in the internal space-time structure of the occurrence itself. In this way, what was once an object of observation in a given time and space in classical physics became the creator of this time and space.

One theory that followed this approach was the organic theory of nature developed by English mathematician and philosopher Alfred North Whitehead. By criticising the materialist view of the universe and any form of subjectivism in the interpretation of natural processes, Whitehead introduced into the discussion an abstract category of an "organism" or "event" uniting in itself the characteristics of matter and spirit.

In concluding that the theory of relativity joined time and space together more closely than previously believed, Whitehead pointed out the possibilities of alternate meanings of time and space could be grasped only by changing science's understanding of these concepts and not simply by creating abstract representations of the existential characteristics of a certain occurrence. Whitehead began by analysing the space-time structure of an individual event or occurrence, and, from this perspective, its relationship with other events. By synthesising findings of such analyses, he tried to define the very category of time.

In analysing the space-time structure of an individual event, Whitehead proposed that such an event, together with a set of all its elements or sub-events, contained two factors:

- The main substantial activity of individualisation when the substance of the main event was disaggregated into sub-events.
- A pattern of "forms" in space and time united through this individualised activity, such as the pattern of connections of sub-events entering into the essence of a given event.

According to Whitehead, this pattern was crucial to understanding the nature and character of time: if the space-time pattern continued through the successive parts of an event and manifested itself in the totality of such an event, it is permanent[8] even as the event itself represented a "life book" of this pattern. The pattern's permanence meant the pattern of one event was also manifested in the scope of its parts, which were separated according to a certain rule. It also implied that any given part of a complete event could not produce a pattern identical to the one that was the result of the whole event.

After developing this definition of the space-time structure of an individual event, Whitehead continued to analyse relationships and interaction among multiple events. He proposed that the mutual influence of events could be explained by assuming that the meanings of space and time for objects in the state of relative motion towards one another are not identical and the variance depends on the duration of the event.

Moreover, if the two objects did not move in relation to each other, they applied the same meanings of space and time to express their durability, but if one object moved relative to the other, their experience of space and time would be different. From here, Whitehead concluded that simultaneity of events existed only if the same temporal system of existence applied to these events, while in other temporal systems these two events would not be simultaneous, even though they might completely overlap. As a result, one event could precede another without special qualifications, if the order occurred in every temporal system.

Underpinning Whitehead's understanding of time was the idea that the connection between events emerged exclusively from the internal, functional relationships. As such, the space-time correlation was an internal component, rather than an external component, and also a condition of existence and relation between events. This internal correlation explained why one event may be found only where it was and as it was and in only one specific set of relations. Since every relation was an inherent part of a given event, this event would not be the same outside that relation.

Whitehead continued, deriving a final conclusion on the nature and character of time. He believed the realisation of events exerted a pattern that required time flow defined by the specific meaning of simultaneity, which linked the given pattern with the specific space-time system. The time flow or a duration contained in the pattern included the range of time, not only an instant, while “a duration, as the field of the pattern realised in the actualisation of one of its contained events, is an epoch.”[9] Whitehead saw time as a sequence of duration of an epoch, whereby “the epochal duration is not realised via its successive divisible parts, but is given with its parts.”[10]

These two ideas—the epochal duration and the claim that the main unit of time is an epoch rather than an instant—necessarily changed the approach to the issues of extensibility and divisibility of time. While Newton started from the belief that a temporal moment was deprived of any possibility of division and extensibility, Whitehead separated the category of time completely from the classical discussion on the extensibility and divisibility by claiming that the entities that follow one after the other were durations that contained the traits of divisibility and extensibility. Pointing to Immanuel Kant’s opinion that points and moments were only boundaries of the continuity of space and time where space contained only space and time contained nothing but time, Whitehead said that because the pattern required a certain duration that must be manifested as an epochal entirety, time represented a sequence of elements that are themselves divisible and real. Further, by becoming temporal, a single duration imposes actualisation with regard to a permanent object: “Temporalisation is not another continuous process. It is an atomic succession. Thus time is atomic (i.e. epochal), though what is temporalised is divisible.”[11]

In the first half of the twentieth century, roughly contemporary to Whitehead’s writings, quantum theory and quantum mechanics were rapidly gaining traction among theoretical physicists. According to this theory, the relations between objects, space, and time at the quantum scale acquired different characteristics than those described by either the Newton or relativist theories of the macro world. Beginning with Max Planck’s interpretation of black-body radiation and Einstein’s understanding of light as a beam of particles of specific energy and impulses—which French physicist Louis de Broglie expanded to encompass all matter—global physicists including Niels Bohr, Paul Dirac, Erwin Schrodinger, and Werner Heisenberg have pursued theories in quantum mechanics that have revolutionised our understanding of matter and its existence in space and time.

The foundation of the quantum theory was represented by the two postulates:

- Every state of a system of particles was described by a wave function of independent coordinates of a system of particles and time, which was finite, uninterrupted, and uniform in every point of space. Thereby, the determination of the coordinate points of a particle was not deterministic, but statistical.
- Wave function changed in time according to the Schrödinger equation, which replaced the earlier determinism of relations between the object, time, and space with the statistical causality.

The Danish physicist Bohr concluded that quantum mechanics was not focused necessarily on the characteristics of micro-objects, but rather with their relations with macroscopic occurrences that could be observed directly. From this view, the principle of statistical causality of physical occurrences in the micro-sphere spread to analysis of causal relations in all physical and natural processes, including the analysis of nature and character of time.

* * *

In the journey from Aristotle to today's quantum physicists, our understanding of the nature of time has evolved considerably, moving from a mere chronological sequence to an objective category. The five distinguishing elements or *differentia specifica* in this progression are:

- From Greek philosophic thought through Descartes, the framework for our notion of time was largely based on the perception of the manifested form, but since Newton it focused on the essence of the given phenomenon.
- From the perspective of relations to a certain occurrence or event, Newtonian physics perceived time as an exogenous category, existing independently of this occurrence, while in the relativistic physics time

represented an integral part incorporated into the occurrence and existing within the occurrence.

- According to Newton, time flowed evenly and independently of the local movement of a given occurrence, making it unaffected, while from the relativistic perspective time was inextricably linked to the motion of an occurrence.
- In classical physics, space and time represented two separated categories. In relativistic physics they composed a uniform space-time continuum with causality of relations between the occurrence—space and time—and in quantum physics this causality has assumed an exclusively statistical character.
- In a pure physical analysis of the category of time, the main unit was represented by an indivisible and non-extensive instant, while the primary unit of time in Whitehead's approach was an epoch, for which the extensity and divisibility have no temporal character, but rather were manifested within events temporalised during the epoch.

The final element is especially relevant since discussions on the temporal dimension of a universal event or process originate within the framework of perceiving time as an objective category. Although Whitehead considered the "pattern of forms" primarily in the context of physical phenomena and derived from this the "atomic structure" of time, his approach could also apply to non-physical events and processes. And within this framework, given specific conditions, it can also be used in the analysis of economic processes. An essential question to consider is whether these processes take place while being uniformly or exogenously, as determined by the calendar presentation of absolute time, or whether they are temporalised and present an "internal time" of their existence.

Notes

1. Isaac Newton, *Philosophiae Naturalis Principia Mathematica*, Bk. 1 (1689); trans. Andrew Motte (1729), rev. Florian Cajori (Berkeley: University of California Press, 1934) 6.
2. Ibid., 7–8.

3. Commenting on Newton's theory, Einstein concluded that absolute time and space do not imply only "physically real," but also "independent in their physical properties, containing physical elements, but not (space and time) affected by physical conditions." Albert Einstein, *The Meaning Of Relativity*, 3rd ed. (New Jersey: Princeton University Press, 1950).
4. Roger Joseph Boscovich, *A Theory of Natural Philosophy* (Chicago and London: Open Court, 1922). According to Boscovich, real space and time cannot be directly cognised by using senses, and the local modes of existence cannot be distinguished in time and space because "our ideas" or perceptions always showed the difference between the new state and the previous one, and not an absolute change, which is not within reach of our senses.
5. Ibid., 407.
6. The relative velocity is not negligible in relation to the speed of light.
7. As presented in V.A. Ugarov, *Special Theory of Relativity* (Moscow: Mir Publishers, 1979) 126–128. In Euclid's space, for example, in Descartes' coordinate system, the square of the distance between the two points is determined by the Pythagorean theorem according to which:

$$\mathrm{d}s^2 = \mathrm{d}x^2 + \mathrm{d}y^2 + \mathrm{d}z^2$$

$$\left(\text{where } \mathrm{d}x^2 = (x_2 - x_1)^2;\ \mathrm{d}y^2 = (y_2 - y_1)^2;\ \mathrm{d}z^2 = (z_2 - z_1)^2\right)$$

In Minkowski's system:

$$\mathrm{d}s^2 = \tau^2 - \mathrm{d}x^2 - \mathrm{d}y^2 - \mathrm{d}z^2 \quad (\text{where } \tau = ct)$$

In such a system there is a uniform manner to establish the temporal and spatial distance between events, which both appear as definitive points each with a specific position in space and an instant in time. Here, the temporal-spatial succession constitutes a "world line" of an object which the events refer to and which is (viewed in the coordinate system x, τ) parallel with the "temporal" τ axis if an object is located at a fixed point on the spatial x axis, that is forming with it an angle Θ (Θ = arctan v/c), if the object moves along the x axis at the constant velocity v. If the event in relation to which we measure the temporal-spatial distance of other events is assumed to be "located" at the origin of the coordinate system ($x_1 = 0$, $\tau_1 = 0$), any other event taking place on the x axis before or after the first event will be visualised by a point in the system (x, τ), with the square of the distance in relation to the origin (the first event) expressed by the following definition

$$s_{1,2}^2 = (\tau_2 - \tau_1)^2 - (x_2 - x_1)^2 = \tau^2 - x^2.$$

Straight lines (world lines of light beams) satisfying the condition, $\tau^2 - x^2 = 0$, that is $\tau = x(ct = xi - ct = x)$, and on which the distance between event—by definition—equals zero, divide the system in four quadrants. Quadrants I and II signify the temporal distance of the certain event located in the origin, while quadrants III and IV signify their spatial distance. Thereby, since in the first quadrant $s^2 > 0$ and $\tau > 0$, the events located in it absolutely follow after the event located in the origin; that is, it signifies an absolute future, unlike the second quadrant where $s^2 > 0$ but $\tau < 0$ which thus indicates an absolute past. In contrast, in the third and the fourth quadrants $s^2 < 0$, and as a result, they signify an absolute spatial distance of the event (from the event located in the origin), provided that this event might take place before, after, or simultaneously with the event located at the origin. So the terms "simultaneously," "before," and "after" are relative for the events located in these quadrants.

Apart from this, in every quadrant there is an equilateral hyperbola which signifies the equidistant points in relation to the origin of the reference frame, with a shape defined by the relation of $s^2 = \tau^2 - x^2$, so that in the reference frame (x, τ) there are two pairs of equilateral hyperbolas: $\tau^2 - x^2 = 1$ (in quadrants I and II) and $x^2 - \tau^2 = 1$ (in quadrants III and IV).

Relative simultaneity of events, time dilation between the inertial reference frames moving relative to each other at the relative velocity v, in Minkowski's four-dimensional temporal-spatial system is expressed as:

- All the events in the reference frame K with the "world line" parallel with x axis (i.e. for which τ = constant) are simultaneous, the same as all the events in the reference frame K', with the "world line" parallel with x' axis (i.e. for which τ' = constant) are also simultaneous.

- Relative motion of the two reference frames causes, in accordance with the Lorentz transformations, an inclination between the axis τ and τ' and axes x i x' (in relation to the common origin of these systems in which $\tau = \tau' = 0$ and $x = x' = 0$), the size of which is determined by the angle $\Theta = \arctan v/c$.
- Distortion of parallelism of the coordinate axis between frames K and K' destroys the absolute simultaneity of the events viewed from these systems, thus causing the time dilation (and, by analogy, the length contraction):

$$\Delta t = \frac{\Delta t_0}{\sqrt{1 - v^2 / c^2}}$$

8. For Whitehead duration meant "repetition of a given pattern in successive events."
9. Alfred North Whitehead, *Science and the Modern World* (New York: Macmillan Co., 1925) 125.
10. Ibid., 125.
11. Ibid., 159.

References

Aristotle. 1952. *Physics*. New York: Oxford University Press.

Boscovich, Roger Joseph. 1922. *A Theory of Natural Philosophy*. Chicago and London: Open Court.

Einstein, Albert. 1950. *The Meaning of Relativity*. New Jersey: Princeton University Press.

Newton, Isaac. 1962. *Philosophiae Naturalis Principia Mathematica*. Berkeley: University of California Press.

Platon. 1999. *Timaeus*. Newburyport, MA: Focus Publishing.

Ugarov, Vladimir A. 1979. *Special Theory of Relativity*. Moscow: Mir Publishers.

Whitehead, Alfred North. 1925. *Science and the Modern World*. New York: Macmillan Co.

3

Time as a Subjective Category

Early theologians were among the first to explore understanding of time as a subjective category, drawing from the work of Alexandrian philosopher Plotinus, and in the seventeenth and eighteenth centuries scientists including Gottfried Leibniz and, later, German idealist philosophers expanded on these concepts, focusing on the rational aspect of such understanding. As it has evolved, the idea of time as a subjective category also became a tenet in modern philosophy.

The perception of time as a subjective category can be traced back to Plotinus, who lived in third-century Rome. Plotinus took into account Plato's concept of time as reflecting eternity and observed that in contrast to eternity, which is continuous, time expresses an actual instant of a specific occurrence and not its temporal flow. Following this, Plotinus concluded that "the life of eternity … is not made up of many times, but is altogether from the whole of time."[1]

Under the influence of such an understanding of time, eternity, and their mutual relation, in the early fifth century St. Augustine, the Christian theologian and philosopher, proposed the first integral conception of time as a subjective category and as a phenomenon of consciousness. Like Aristotle, Augustine rejected the view that time was simply the

Ž. Rohatinski, *Time and Economics*, DOI 10.1007/978-3-319-61705-3_3

movement of celestial bodies, arguing, "Or, if the lights of heaven should cease, and a potter's wheel run round, should there be no time by which we might measure those whirlings, and say, that either it moved with equal pauses, or if it turned sometimes slower, otherwhiles quicker, that some rounds were longer, other shorter?"[2]

Augustine concluded there was a significant difference between the motion of a body and the measure of this motion. Although a body at times moved and at times remained still, time was used to measure not only its motion, but also its rest. Following that, Augustine defined the essence of the problem by questioning whether, if the motion of a body is measured by time, the time itself is measured as well.

Augustine found a solution to the problem in the view that the measurement of time—and thus time itself—represented a significantly different essence in relation to the physical movement of a body and the processes in nature. Time, as such, belonged to a phenomenon of consciousness or soul:

> It is in thee, my mind, that I measure times. Interrupt me not, that is, interrupt not thyself with the tumults of thy impressions. In thee I measure times; the impression, which things as they pass by cause in thee, remains even when they are gone; this it is which still present, I measure, not the things which pass by to make this impression. This I measure, when I measure times. Either then this is time, or I do not measure times.[3]

As a result, time represented a "distension of a soul," which Augustine also used to explain the flow of time as different states of consciousness.

Extrapolating such a concept of time from its theological elements suggests an analogy in approach between the notion of time as a subjective and an objective-absolute category, which is reflected in two elements:

- In the context of the human cognition, both approaches take time, as well as eternity or never-ending time, as an objectively given category originating together with the universe, independent of the process contained therein and in the human consciousness.

- Absolute time is relatively manifested through the motion of celestial bodies, which represents an objective conception, and through subjective cognition, which belongs to a subjective conception.

This analogy implies that an understanding of time as a subjective category does not deny its objective existence as an absolute category, but claims that the *real* time originates in joining its objective existence expressed by change or motion of the occurrence and a subjective cognition of this change that contains the elements of memory and expectance, which happen only in an actual instant of the occurrence or event. Ignoring for the moment the subsequent modifications of this approach, the initial basis and conclusions of such a conception, especially within the framework of a transcendental approach, remain present in modern philosophy.

The first attempt to describe the scientific foundation and provide a conceptual basis for this idea was by Leibniz (1980), who perceived an "objective-idealistic" image of the world. His theory was initiated by criticism of Descartes' mechanistic theory of substance, which was separated and passive in the context of the role and character of consciousness. In contrast, Leibniz proposed the theory of monads, simple substances capable of action that represent individual, indivisible, non-extensible, active, and self-active "spiritual" atoms. Monads were an integral part of all occurrences in nature and represent a uniform foundation, a principle of life, a creative condition, and a primal force of the entire universe. Although, analogous to the notion of consciousness, monads represented a "spiritual" category, and Leibniz perceived them as inseparably related with matter. Also counter to Descartes, in Leibniz's view matter was not the opposite of consciousness or spirit, but was perceived as a spiritualised activity. In that way, Leibniz defined the unity of matter and consciousness.

At the same time, Leibniz pointed out that each monad possessed the properties of perception and appetitions. According to him, the perception should be sought in a simple substance, while appetitions implied a tendency to change from one mental state to another.

The relations between monads were regulated by the two main principles of nature:

- Continuity or quantitative relations, in which occurrences in nature were not taking place in rapid sequences, but constantly and gradually.
- Identity or qualitative relations, in which each individual monad was different from other monads according to its qualitative characteristics.

Considering that a monad was indivisible by definition, according to Leibniz, it was different than any other monad only because of its internal properties relating to perception and desire. Further, Leibniz divided monads according to "clarity of perception," ranging from unconscious perception (matter) to absolute conscious perception (God). In terms of human consciousness, a monad acquired a new qualitative property of perception, which was self-consciousness or a reflex cognition of an internal state, and therefore a perception of the monad.

By defining the relation between consciousness and self-consciousness as a basis of perception and cognition of the totality of occurrences and processes in nature and also starting from the principle of continuity of these processes, Leibniz also built a theory of time as a subjective category. He concluded that the change of occurrences in nature defined the succession of time as an important category of development and existence of these occurrences. He perceived the concept of time through the decomposition and succession of changes that could happen within time or, more simply, that time consisted of successive changes. Because under the principle of continuity everything in the universe was connected, the present was always burdened with the future and no given state in nature could be explained except in terms of the immediately preceding state. Here, the monad, with its property of perception, represented the whole of the prior, actual, and future occurrences. Apperception presented the self-consciousness transfer of this perceived "external" time flow of the occurrence into the internal "spiritual" event and initiated the "real" time or subjective temporalisation of the external time flow of the occurrence.

By creating this framework, Leibniz rejected Newton's understanding of time as an exclusively objective category and set the basis for a subjective conception of time, which became one of the central parts of the idealist theory of the totality of occurrences and processes in nature and society within the philosophy of nature and eighteenth-century German idealist philosophy. Although differences in the notion occurred among the movement's various authors, such as Immanuel Kant, Johann Fichte, Georg Hegel, and others, its essence was the most thoroughly defined by Kant in *The Critique of Pure Reason.*

In developing the theory of transcendental aesthetics, Kant denied any form of absolute reality to time: time was independent of the form of our inner sense and attached to things absolutely as a condition or property of the certain appearance. This also introduced the concept of the "transcendental idealism of time." Here, time and space represented pure or absolute sensation. They created a form in which appearance was sensed and joined within themselves the subjectivity of sensation and experience or the self-consciousness of this subjectivity.

In addition, a framework was created describing the two sources of all man's knowledge of appearances, events, and the interrelations of nature and society. These two sources are receptive sensibility or the ability to sense "outer" appearances and reason or an awareness of this sensibility.

From here, Kant set the three main postulates on the role and character of time as a subjective category, the second and third following from the first:[4]

- Time does not exist for itself or exist within things as an objective property, and it does not persist when abstracted from all subjective conditions of sensible intuition. On the contrary, "time is nothing but a subjective condition under which alone intuition can take place in us."
- "Time is nothing but a form of inner sense, that is intuition of ourselves and our inner state, and it determines the relation of representations in our inner state."
- "Time is the formal *a priori* condition of all appearances. Since all representations, whether they have for their object outer things or not, belong in themselves, as determinations of the mind, to our inner state, and since the inner state stands under the formal condition of inner intuition and so belongs to time, time is an *a priori* condition of all appearances."

Kant's understanding of nature and the character of time was also greatly contested within the criticism of the main principles of the German philosophy offered by the classics of dialectic materialism, as well as by the authors of other ideological viewpoints including Friedrich Nietzsche. This criticism has inspired a group of modern philosophers, especially representatives of the phenomenology movement, to reinterpret and extend Kant's ideas to re-establish the concept of time as a subjective category. In this context, the works of the German philosophers Edmund Husserl and Martin Heidegger bear great significance.

Starting from the Kant's postulate that time was a form of inner sense and as such represented a universal and pure intuition and taking into account that what the observer initially has in mind in experience of all appearances was their pure succession, Heidegger (1968) pointed out that Kant does not completely deny the temporal dimension to appearances in nature, but treats time as a characteristic indirectly arising from the observer's perception.

By qualitatively separating the "real" observer's time (subjective) from the "mechanical" or calendar time in which the appearance occurs, Heidegger continued to analyse the notion of real time and how it forms. He started by postulating that every sensible intuition originated from the so-called transcendental imagination. This imagination formed the essential unity of pure intuition (time) and pure thinking (apperception) and synthesised within itself the following elements:

- Faculty of imaging, as a representation of the present time.
- Faculty of reproduction, as a representation of the past time.
- Faculty of anticipation, as a representation of the future time.

In this way, the category of time itself as a pure intuition was based on the transcendental imagination.

As a result, time constantly flowed as a pure succession of the series of "now" moments. This succession was sensed by using pure intuition without objectifying it. However, Heidegger concluded that pure intuition could form the pure succession of the series of "nows" only if it in itself was an imagination that images, reproduces, and anticipates. Therefore time, within the horizon in which we "calculate with time," must be taken as the series of "nows," but this sequence did not represent time in its originality. On the contrary, the transcendental imagination was what allows time to flow as a succession of "nows" and was itself truly primordial time.[5]

Going further, time as a pure intuition was neither that which was sensed by pure intuition nor an intuition missing an "object." Time as a pure intuition was also an intuition that formed its own object. This pure intuition induced itself by its object, which it formed without the help of experience, so by its very nature time self-induces.[6]

These ideas, first formulated in 1927 in Heidegger's *Being and Time*, significantly influenced the approach and analysis of the category of time within modern philosophy. Some authors accepted and elaborated upon them, while others contested them, in parts or as a whole.

Among the first to accept these concepts, French philosopher Maurice Merleau-Ponty deserves special attention. By integrating positions held by Heidegger and Husserl, Merleau-Ponty formulated a conception of time as a transcendental horizon uniting the observer and the being and constituting a primordial or authentic time of the being itself.

Merleau-Ponty began a discussion on the nature and character of time by stating that one could not accept a conception in which time in nature objectively originated from the past and flowed towards the present and future, in the process making the present the consequence of the past and the future the consequence of the present. He confronted such a conception by asserting that in nature there are no successive appearances in terms of time, but "if we separate the objective world from the finite perspectives which open upon it, and posit it in itself, we find everywhere in it only so many instances of 'now'."[7]

Merleau-Ponty also observed that since such an objective present has no temporal character, not only was it wrong to see time as a "succession of instances of 'now' " because this treats the past and the future as presents, it was also inconsistent because it destroyed the notions of "now" and succession. As a consequence, the successive development of the appearance, process, or the event that took place in a specific "time" range with the characteristics of the past, present, and future had no place in the objective world. Instead, it was created by a finite observer from the spatial-temporal totality of the objective world. This meant that time was "not a real process, not an actual succession that I am content to record."[8] Time originated from the observer's relation towards things and in things themselves, while the future and the past essentially were a permanent presupposition and a permanent outliving. As such, past and future were only constructs in the present of an individual observer and did not exist outside the present.

Such real time, in which the observer internalised the changes in nature and their passing, was constituted by expanding the consciousness. In that sense, the present with its "indelible individuality" represented the present only through relations that "the consciousness positions between it, the past, and the future." Going further, Merleau-Ponty claimed that the past, present, and future dimensions of time were not presented through discrete distinguishing actions and the observer was not passing through a sequence of "nows" organised as a timeline, but rather the passing moment was modified in every oncoming moment.

Merleau-Ponty elaborated this thesis further by reinterpreting Husserl's phenomenological time scheme, which represented the current cross section of an "objective" time by a series of non-successive instances of objective "nows" that emerged from one another and were projected or profiled onto the observer's consciousness and the observer's "now." Through transformation of objective instances into the observer's "now," the observer subjectively perceived an objective instance, which moved from the observer's present to the observer's past. In this way, the observer created an awareness of the past and dimensions of time were formed. On this basis, Merleau-Ponty concluded that time was not a line following an objective course, but a measure of intentionality and subjective perception of the objective "nows."

However, since the observer was also positioned in the temporal-spatial totality of the objective world—that is, existed in the objective "now"—the observer's perception of time was not to imagine the transfer from one present into the other, but the actualisation of this transfer. This implied that "time is not an object of our knowledge, but a dimension of our life." For Merleau-Ponty, time was not an external dimension or coordinate of an appearance or event, but the appearance itself. Finally, he said the "temporalisation" of an objective event was not successive since "the future is not posterior to the past or the past anterior to the present. Temporality temporalises itself as future, which lapses into the past by coming into the present."[9]

Such conclusions essentially conform to Kant's theoretical foundation, which held "up to now … it has been assumed that all our cognition must conform to the objects.… Let us once … assume that the objects must conform to our cognition." But neo-Marxist philosophy rejected this phenomenological conception of time as a subjective category almost entirely. Yet this refutation led neo-Marxists only a partial return to Newton's original theory of absolute, objective time. It questioned Newton's concept of non-afflicted time, replacing it with the concept of differentiated time, dependent on the content and "dynamic" of changes occurring within this time.

German Marxist Ernst Bloch in his last book, *Experimentum Mundi*, pointed out that time as a whole did not overlap with a linear succession of the time itself as measured by the clock. He emphasised that time was independent of people and of unclosed contents occurring within it. Although Bloch did not deny the existence of absolute quantum of time as measured by the clock, he asserted that the time frame is affected by the content occurring in different forms of time. This meant that the flow of time is differentiated to various levels that do not allow the uniform antecedency and subsequence of a formally homogenous composed sequence leading from the past into the future. From this, even different quanta of time were not homogenous because they were mostly qualitatively different among themselves.

Bloch continued that since time is related to breath, its speed and density were always different. Essentially, there was more time or less time depending on the size of change within its frame. Every living being had its own time, depending on the current stage of its life, and this time lagged behind or proceeded the uniform time measured by the clock. This dissimilarity of time was essentially negated in the abstract of time measured by the clock, which insisted of permanence and uniformity, and a metric order, although unreal, was unavoidable for practical reasons. However, regardless of where it appeared and was being continuously and materially condensed in formal time as measured by the clock, different groups of time appeared repeatedly.

Within this context, Bloch concluded that time was not an abstract scheme of change, but a concrete-elastic field of its course, changing in parallel with the type and the content of the change itself. Time was neither a subjectively ideal illusion nor a hypostatised specificity that spread out by itself with change occurring as though it were in some pre-arranged tube that channelled it homogenously. He believed the hypostasis of the uniformity of time—represented as objective and appearing also in uniform succession in purely mechanical materialism—was as untenable as was contesting the real character of time within subjective idealism. Time existed exclusively in material change, with material change as a mode of its course, forming and changing with it.

From this understanding of the character of time, Bloch inferred that an instant of "now" was too short to retain anything, and so an instant was something in which nothing has yet to occur, nothing continues, and nothing is gathered. Only after the "now" has passed and moved away from itself could it be experienced and become "aware" and "given" by itself. Since a "now" contained no formation and destruction of an appearance, "mere 'now' is groping in the dark," Bloch said, and so the succession of a sequence of "nows" cannot form a continuous flow of time. On the contrary, every "now" is repeatedly interrupted by a single "never," which, only in their integrity, result in a time shift and the "creation" of the present as an experienced, continuous, and relatively broad temporal category.

Continuing this logic, the present represented a mixture of "now, the past, and the future." The ordinary sequence and order of the temporal modes—past-present-future—would have to be changed. From this perspective, according to Bloch, only "now" itself is the beginning, and its proximity to the living being, without deviation from itself, represented the beginning everywhere.

Taken as a whole, conceptualised time as a subjective category raised a series of important questions. Augustine essentially defined the first, which centres on the course of time and its measurement. From this one can ask: Does chronological or calendar time defined by the motion of celestial bodies represent only one potential conventional, collective, or unique external measure of the duration of a process or event, but not the time in which this process itself occurs? If so, every individual, objective process has its own internal, specific duration. Also, each process has its internal time perceived as a succession and duration of the change, which depends on its character and property, and this is measured by its specific internal measuring system, independent of any external measuring system. In other words, internal time remains unchanged regardless of the possible changes in the external measuring system.

This hypothesis would suggest that this time is objective. If the analysis is extended to two or more processes, then relativistic relations are established between the internal times of these processes. The internal time of one process observed from the aspect of the internal time of another process is transformed into the measuring system of the other process. Within this transformation, a relation is formed, focused on whether an occurrence is faster or slower or whether internal time flow is faster or slower. Naturally, the objectivity of this transformed time cannot be denied, but it is not original and represents a relativistic derivative.

The second question reflects the essence of the conception of time as a subjective category and relates to the understanding and the idea of the flow of time in the human consciousness. If we assume that the imaginary, unconscious and, in the final consequence, timeless instance of "now" is an origin of the course of time, can the recollection of past and expectation of future together with this instant create a "temporal consciousness" and can the idea of a temporal course of events be formed in the human consciousness? Such recollection is not only subjective, but also static, and it is not necessarily connected to the chronological or calendar time. Recollection is simply an idea of the sequence of occurrences, external or those within the consciousness, while an idea of the duration of the event or process is created by comparison with another corresponding process or event, usually in terms of chronological or calendar time.

At the same time, a living being also exists in the objective time, and its course defines an object or "life process." This objective time and the processes that populate it "fulfil" the human subjective idea of the course of time, and they also affect it and model it. Therefore, the dynamics of the objective time and its flow create the dynamics of the subjective idea of time.

Finally, the third question centres on finding a "uniform" time and corresponding course. Do the differentiated times of processes, their differentiated measuring systems, and the subjective idea of the course of time result in a multitude of temporal modes and dynamics? Under these conditions, the notion of a common measurement seems elusive, but without one the issue of the duration of the processes and their simultaneity would remain unresolved. Perhaps, different courses of time and the subjective idea of them are homogenised at a high level into an abstract chronological time. For such a time, whether it is relatively manifested by the motion of celestial bodies or in some other way is negligible. Within such an abstract time, an "obvious" simultaneity or non-simultaneity of events and processes is formed, as well as an image of their identical or non-identical duration. Relations established in this manner are also abstract and do not reflect the functional temporal relations between the objective events and processes, even if such relations exist in terms of content.

* * *

Newton's concept of absolute time fared badly in subsequent centuries. Later scientists and philosophers would be divided into those who saw time as an objective category and who saw it as a subjective category, but despite differences in approach, methodology and analysis, conclusions, and the theoretical and practical implications of their ideas, they all rejected the absolute nature of time in favour of relativistic approaches.

Still, the possibilities for applying such diverse understandings of the category of time diverged significantly in the actual research across various scientific fields. In physics, the theory of relativity of time was needed to define and analyse certain processes, for instance, while the application of relativistic concepts time, either in definition and analysis, in many social sciences, including economics, seemed not only unnecessary, but almost completely unexplored. The problem was obviously aggravated by the character of research in social sciences. The possibilities are relatively limited to analyse social processes and empirically prove or disprove a certain conception of time in which these processes occur. Everything appeared to be reduced to logical constructions and hypotheses.

Bloch's concept of the differentiated density of time is helpful. From the perspective of everyday human life and activity, it integrates the two "obvious" aspects:

- Uniform chronological flow of (absolute) time.
- Differing "densities" of this time, dependent on the processes occurring within it and different levels of "crowdedness" of chronological time with processes and events.

It remains unclear whether these "obvious" aspects of time can completely define the nature and character of the temporal dimension of the social events or the processes formed in their interactions or whether another conception of time would be better.

From the perspective of economics—without any claim of wider applicability—four hypotheses emerge that may contribute to the understanding of the concept of time in this field:

- Absolute time passing evenly from the point of chronology is merely an abstraction in which the objective, specific-functional times of different processes are homogenised in a non-functional manner.
- Each individual process exists in its internal, functional time that is specific and independent from any possible changes in the external time measurement system.
- Internal time measurement depends on the objective properties of a process, while the integration of time flow and its measuring system makes the process "temporal."
- The subjective cognition of time represents a "static" awareness of the sequence of the processes and their duration, which is "becoming dynamic" because of the existence of an observer in the objective time.

Notes

1. Plotinus, *Eneade* (Beograd: Grafos, 1982) 25.
2. Augustine, *The Confessions of Saint Augustine* (Grand Rapids, MI: Christian Classics Ethereal Library) 286.

3. Ibid., 290.
4. Immanuel Kant, *The Critique of Pure Reason* (London: Palgrave Macmillan UK, 2003) 76–77.
5. Ibid., 113–114.
6. Ibid., 121.
7. Maurice Marleau-Ponty, *Phenomenologie de la Perception* (Taylor and Francis e-Library, 2005) 479.
8. Ibid., 478.
9. Ibid., 488.

References

Augustine. n.d. *The Confessions of Saint Augustine*. Grand Rapids, MI: Christian Classics Ethereal Library. www.ccel.org

Bloch, Ernst. 1980. *Experimentum Mundi*. Beograd: Nolit.

Heidegger, Martin. 1968. *Kant and the Problem of Metaphysics*. Bloomington: Indiana University Press.

Kant, Immanuel. 2003. *The Critique of Pure Reason*. Translated by Norman Kemp Smith. London: Palgrave Macmillan UK.

Leibnitz, Gottfried Wilhelm. 1980. *Izabrani filozofski spisi*. Zagreb: Naprijed.

Marleau-Ponty, Maurice. 2005. *Phenomenologie de la Perception*. Translated by Colin Smith. London and New York: Taylor and Francis e-Library.

Plotinus. 1982. *Eneade*. Translated by Slobodan M. Blagojević. Beograd: Grafos.

Part II

The Structure and Character of Economic Activity

Before considering how a modern concept of time should influence our understanding of economics, we must first reach a common view of the structure and character of economic activity. In particular, we should explore ideas that define the properties of conceptual categories relevant for new insights into "economic time."

The main elements needed for this perspective are:

- The event or process for which the time dimension is being analysed.
- The rules under which this event or process unfolds and which defines its internal measurement of time.
- How humans perceive these events or processes and rules and how this perception defines observed sequence and length, as well as the subjective awareness of time.

An understanding of each of these elements is built around a series of explorations, from the historical character of economic activity to the existing concepts of economic modelling.

4

Historical Character

The notion of "economic activity" is amazingly fluid. How it is defined can depend largely on whether it is being viewed from the perspective of a specific scientific discipline, religious doctrine, or political or social ideology. For example, in 2005 the US Supreme Court said economic activity comprises the "production, distribution, and consumption of commodities," a formulation—taken from a standard dictionary—that dissenting justices found to be too broad and potentially encompassing "all of productive human activity."[1] A psychologist, on the other hand, could see this definition as too narrow because it doesn't include the motivational and behavioural dimensions behind production, distribution, and consumption.

In essence, thinking around economic activity coalesced around the essential influence it exerts on the development of the society and relations between its subjects. The ideological basis of this stand was formulated by Karl Marx in the preface to *A Contribution to the Critique of Political Economy.*

Marx's long-time partner Friedrich Engels added in an 1894 letter:

> Political, juridical, philosophical, religious, literary, artistic etc., development is based on economic development. But all these react upon one

Ž. Rohatinski, *Time and Economics*, DOI 10.1007/978-3-319-61705-3_4

> another and also upon the economic base. It is not that the economic position is the cause and alone active, while everything else only has a passive effect. There is, rather, interaction on the basis of the economic necessity, which ultimately always asserts itself.[2]

The relationship, however, should not be interpreted as deterministic. The issue of dominance is one of the fundamental questions facing almost any field of humanities, whether economic activity, some other social or natural activity, or a general human awareness on the totality of these.[3]

In the narrower sense of research in economics, various influential writers have attempted to define the scope of economic activity, including:

- *Engels*: A unity of production and exchange of the material means of subsistence in human society.
- *Alfred Marshall*: Man's conduct in the business part of his life.
- *Lionel Robbins*: Human behaviour as a relationship between ends and scarce means that have alternative uses.
- *Paul Samuelson*: Making a choice, with or without the use of money, to employ scarce productive resources, which could have alternative purposes, to produce various commodities over time and distribute them for consumption, now and in the future, amongst various people and groups of society.
- *Oscar R. Lange*: Production and distribution of goods representing two social areas of human activity.

The differences in these definitions can be largely attributed to disparate starting points and ideologies. But despite these differences, a general consensus can be described concerning two central aspects of economic activity.

First, economic activity is the social activity that establishes specific relationships between the subjects in society and nature through the production factors, and, at the same time, among the subjects[4] in society through the relations of production. From this, economic activity is multi-valued and complex and includes a relatively broad range of material and social aspects and contents.

And second, the social character of economic activity implies that it has a governable character, which allows an individual and a society to directly or indirectly influence its content, form, course, and characteristics within objective natural, material-technical, and wider social constraints.

Drawing from these two aspects of economic activity, the scope and intensity of any potential deliberate social impact on economic activity and the relations between the subjective and objective sphere of that activity comes to the fore. In the early twentieth century, German sociologist and economist Leopold von Wiese constructed a formula that suggested a social process was the result of an individual's behaviour and the prevalent social situation.[5] Although the formula itself did not apply directly to an understanding of economic activity, because of its material and social complexity, some of the characteristics described in explaining the formula are relevant from the perspective of this book. In von Wiese's view, the contents and nature of these characteristics depended largely on the economic structure of a society and the level and form of its economic organisation, with completely non-hierarchical and completely hierarchical social structures as the two extreme or pure cases.

The hierarchic character of a system is a complex notion with different definitions in the literature of economics and other social sciences. The approach of Hungarian economist and academic János Kornai is rather indicative of this.[6] By dividing economic subjects or organisations into real subjects and control subjects, he distinguished the two types of hierarchic characters, that is, two types of subordination or superiority of subjects, in the economic system:

- Subordination and superiority based on the right of issuing commands or directives, which are subject to legal sanctions.
- Subordination and superiority based on the monopoly of unavoidable information.

If the economic structure of a society were viewed as a non-hierarchical system and the main relationships between the subjects of economic activity horizontal, which historically corresponds with the stage of trades in the late feudalism and early stages of capitalism, the main characteristics of economic activity would be:

- The freedom and intensity of the conscious impact of each individual subject on its economic activity are negatively correlated to the number of other subjects affecting this activity. This implies that expansion of the conscious action from one to more social subjects and, as a result, the expansion of overall economic activity lower the level of freedom and intensity of the partial impact an individual subject has on the common activity.
- In contrast to the reduction of individual freedom and intensity of the conscious impact made on economic activity, a greater number of social subjects affecting the economic activity generally increases the total effect of such an impact.
- There is usually a much bigger possibility of conscious influence on initiating the contents and the course of economic activity than on changing the consequences of the "initial" conscious activity.[7]

Classical political economics illustrated these characteristics as a market with a large number of "atomised" producers and consumers of a certain product or substitute. Each of these subjects, within some objective frameworks, had a relatively broad range of independent actions, such as changing the scope of production, focusing on the production of another product, switching consumption to a substitute, or changing the scope of consumption, among others. No single producer or consumer was powerful enough to influence total supply and demand.

The conflict among these unilateral decisions and actions, each just a small portion of the overall dynamic, creates competition and increasing uncertainty about the effect of each individual activity. By grouping these economic subjects into associations such as gilds—usually to reduce competition among them—and keeping a non-hierarchical economic structure still intact, an individual's freedom of activity is reduced because it is then partially subjected to rules imposed by the association. On the other hand, the association's aggregated influence on total supply, demand, or both likely increases.

By consciously influencing the scope and structure of production and supply, these subjects as an association initiate a course of economic activity in which they exercise a greater level of freedom and efficiency of action than they would by trying to influence the consequences of such an activity, for example the reaction of consumers or the prices of products. Of course, there are also ripple effects, as well as the influence consumer behaviour and overall price trends had on decisions related to production and supply.

If the concentration and centralisation of capital and increased economic strength exercised by a state or quasi-state institution become the predominant characteristic of a contemporary economic system, then a vertical differentiation of the economic structure of a society or a hierarchical system develops.[8] Historically, largely non-hierarchical systems tended to move towards an increasingly pronounced and differentiated hierarchical structure. In such a system, the influence mutual competition, which is the main limiting factor of freedom of individual subjects, exerts on economic activity that was found in a non-hierarchical system is significantly weakened and is much more affected by the elements of domination in the material sphere of this activity. This domination in the material sphere inevitably begets domination in making decisions because the higher the level of development and complexity of an organisation, the more independent its operations are from outside interference.[9]

Considering an economic subject's ability and freedom to influence economic activity, a hierarchical system is based on the principle of prioritising the activities between the levels or sectors, thus establishing hierarchical relations among them. Through its actions, the higher level or sector of the economic system predetermines the activity of lower levels. In other words, the higher level specifies the framework for the independent activity of those lower in the hierarchy, as well as directly influencing their decisions. The transfer succeeds as long as there is an adequate intersectoral exchange of information.

Such an exchange of information plays a role similar to mutual competition based on market information for economic subjects in a non-hierarchical system. However, unlike in a non-hierarchical system, in a hierarchical system the exchange of information between subjects on different levels is direct. Further, in hierarchical systems there are different motives and objectives for economic activities by subjects on different levels, which lead to the formation of hierarchical relations among the objectives of economic activity in a society and not just the actions.

It follows that increased freedom and efficiency of the conscious action at the higher levels of a system is a primary characteristic of the influence exerted on economic activity in a pure hierarchical system. As a result, the potential impact on the consequences subsequent to the initial economic activity is also greater.

In practice, however, the hierarchic component of the system does not suspend entirely an individual subject's activity or influence on the economic activity, which is a hallmark of a non-hierarchical system, although it limits them to relations formed within a given hierarchical level. The relations among the subjects on different levels are dominated by hierarchical subordination, which forms a complex vertical-horizontal structure of possibility, freedom, efficiency, motives, and objectives surrounding a subject's influence on the economic activity. To understand the economic objectives and the means required to meet them within a hierarchical structure, a grasp is needed of one of the defining characteristics of economic activity: the alternative use of scarce productive resources for achieving certain objectives.

Just as social organisation and structure change over time, the concept of economic activity also has a historical character that is shown in the gradual development and change of its content, form, and properties. This process relies on the material and spiritual development of the productive forces of a society and the ethical, political, and social norms. In principle, though not necessarily in parallel, these changes affect all aspects or elements of economic activity: its main agents, driving motives, organisation, technical and technological conditions, the mode of production and exchange, productive relations, the manner of social impact on the course of the certain activity, and others. Essentially, the changes affect the relations between the individual production factors and, in turn, the structure and intensity of the impact of the individual elements on the total economic activity.[10]

These changes, as well as the underlying economic activity, are continuous, but not uniform in intensity or linear, making them dialectic.

Notes

1. Gonzales v. Raich, 545 U.S. 1 (2015).
2. Karl Marx and Frederick Engels, *Collected Works. Vol. 50, Letters, 1892–1895* (London: Lawrence & Wishart, 2010) 265.
3. One prominent example is Max Weber's theory of religion-ethics-legitimacy, which distinguished the four main types of social action: means-end rational, value-rational, traditional, and affectual. The first, means-end rational, represents a rational activity aimed at achieving an individual end by using the means of the certain technical efficiency and the cost of its use and can largely be identified with economic activity. The German sociologist pointed out this type of social action always exists within a legitimate system regulated by legal norms and generally based on the relevant ethical principles. Since the actions are essentially dependent on religious beliefs and norms, Weber emphasised a close relationship between religion and social activities, that is, the influence of the "state of mind" on the material aspect of society (Max Weber, *Die protestantische Ethik und der Geist des Kapitalismus* [Sarajevo: Veselin Masleša, 1968] 206).
4. In classical economics, "subject" usually refers to households, taken as an elementary subject in the process of production. In contrast, within the systems theory, subjects are separated by their traits and the relationships of different complexity are established among them, depending on their structure and properties. In this sense, Hungarian economist and academic János Kornai distinguishes three subjects within an economic system:

    - Economic unit: an "indivisible element of the economic system that behaves according to definite regularities and responses to outside stimuli in a predictive manner."
    - Economic organisation: a "social formation consisting of persons who associate in order to perform some definite social-economic functions."
    - Economic institution: a "subsystem of the economic system, differentiated by its organisational and legal characteristics."

    (János Kornai, *Anti-equilibrium* [Zagreb: Centar za kulturnu djelatnost, 1983] 53, 95, 102).

5. As described by Barnes in the early nineteenth century, in one of the most significant representations of the "formal-rational "approach in sociology, German sociologist and economist Leopold von Wiese presented a formula for identifying the factors in any social process. According to his formula, the social process (*P*) is a result of individual behaviour (*H*) and the prevailing social situation (*S*). Further, he implied that the behaviour of individuals depends on their innate traits (*I*) and earlier experiences (*E*), while the social situation in which the process takes place depends on non-human factors (*U*) and the behaviour of other individuals taking part in the process in any way (*IxE*).His constructions included:

   $P = HS$
   $H = IE$
   $S = U(IxE)'$

   and the total:

   $P = IEU(IxE)'$

   (Harry Elmer Barnes, *An Introduction to the History of Sociology* [Chicago: University of Chicago Press, 1966] and Leopold von Wiese, *Allgemeine Soziologie. Vol. 1, Beziehungs lehre* [Munich: Duncker und Humboldt, 1924]).
6. János Kornai, *Anti-equilibrium* (Zagreb: Centar za kulturnu djelatnost, 1983) 86–90.
7. In practice, the hierarchical character comprises a wider set of direct and indirect relationships between the economic subjects. Apart from the elements analysed by Kornai, these relations also stem from a dominant position of the individual subjects on the market, an ability to influence consumer preferences, a monopoly over a certain technology, and so on. The simplest explanation of a concept of hierarchical character of the economic system is that it represents different levels of ability of an individual economic subject to influence the behaviour of other subjects and the total economic activity.
   When defining the main characteristics of deliberate activity of economic subjects in non-hierarchical systems, as well as the systems themselves, we come across a series of important problems. These problems are difficult to resolve even though they have been thoroughly researched on a theoretical basis. Generally these include:

- Disturbances in the non-hierarchical character of a system because of vertical connections of subjects in the productive process.
- Different economic potential and the total "force" of individual subjccts.
- Significantly restricted availability of individual factors of production.
- Technological monopolies.
- Inelastic demand for the certain products.
- Imperfect availability of market information.
- Restricted mobility of subjects and economic factors.

This indicates that, in practice, neither pure hierarchic nor non-hierarchic systems exist and neither have ever existed, but they have always been present in some combination.(For examples of these problems, please see Joan Robinson, *The Economics of Imperfect Competition* [London: Palgrave Macmillan UK, 1969].)

8. John Kenneth Galbraith, *The New Industrial State* (Zagreb: Stvarnost, 1970) 23–32.
9. John Kenneth Galbraith, *Economics and the Public Purpose* (Rijeka: O. Keršovani, 1979) 47.
10. John Kenneth Galbraith's "marginalist" interpretation states that power belongs to the production factor with the lowest elasticity of demand, on the border of cost-effectiveness. This inelasticity may be the consequence of a shortage, of the efficient control over the supply, or of a combination of these two (John Kennet Galbraith, *The New Industrial State* [Zagreb: Stvarnost, 1970] 66).

References

Barnes, Harry Elmer. 1966. *An Introduction to the History of Sociology*. Chicago: University of Chicago Press.

Galbraith, John Kenneth. 1970. *The New Industrial State*. Zagreb: Stvarnost.

———. 1979. *Economics and the Public Purpose*. Rijeka: O. Keršovani.

Kornai, János. 1983. *Anti-equilibrium*. Zagreb: Centar za kulturnu djelatnost.

Marx, Karl. 1993. *A Contribution to the Critique of Political Economy*. Moscow: Progress Publishers. On-line version: Marx.org.

Marx, Karl, and Frederick Engels. 2010. *Collected Works. Vol. 50, Letters, 1892–1895*. London: Lawrence & Wishart.

Robinson, Joan. 1969. *The Economics of Imperfect Competition*. London: Palgrave Macmillan UK.

Weber, Max. 1968. *Die protestantische Ethik und der Geist des Kapitalismus*. Sarajevo: Veselin Masleša.

Wiese, Leopold von. 1924. *Allgemeine Soziologie. Vol. 1, Beziehungs lehre*. Munich: Duncker und Humboldt.

5

Circular Character

Within the continuous and historical course of dialectics, economic activity is considered to be circular and to contain four phases: production, distribution, exchange, and consumption. This series is successively and repeatedly reproduced either with an unchanged volume, simple reproduction, or a changed volume, extended reproduction or recession.

Already in 1758, French economist François Quesnay was describing the circular character of economic activity in his *Tableau économique*. Based on a physiocratic perspective, which allowed land as the only source of wealth and the agriculture as the only activity multiplying this wealth, Quesnay presented the circular flow of products and money among the three social classes taking part in economic activity: the productive class replenished the annual wealth of nations by cultivating the land; the proprietary class lived off the income or net product of agriculture, which was paid annually by the productive class through leases, and comprised the sovereign, landowners, and collectors of tithes; and the sterile class, comprising all those individuals working in other occupations and whose expanses are covered by the productive class and the proprietary class.[1]

Ž. Rohatinski, *Time and Economics*, DOI 10.1007/978-3-319-61705-3_5

Quesnay's scheme for the circular flow of products and money among these classes was composed of three main elements:

- Gross product of the productive class, valued at five billion livres, of which two billion livres are retained by the productive class and the remaining products, worth three billion livres, are sold to the proprietary and sterile classes.
- Manufactured products of the sterile class, worth two billion livres.
- Money assets totalling two billion livres that are retained by the productive class and represent the net product of this class.

Within this framework, the productive and sterile classes entered the circle of economic activity with the annual advances of two billion livres to the productive class, expenses for agricultural work, and one billion livres to the sterile class, funds for buying products from the productive class.

Under such conditions, the circular flow of economic activity—trade and the transformations of products and money—comprised six rounded transactions, which, based on Marx's reconstruction,[2] can be graphically presented, with the top diagram representing the flow of money and the bottom that of production (Fig. 5.1).

However, even after extrapolating from the initial or physiocratic basis, Quesnay's scheme of the circular flow of economic activity remained

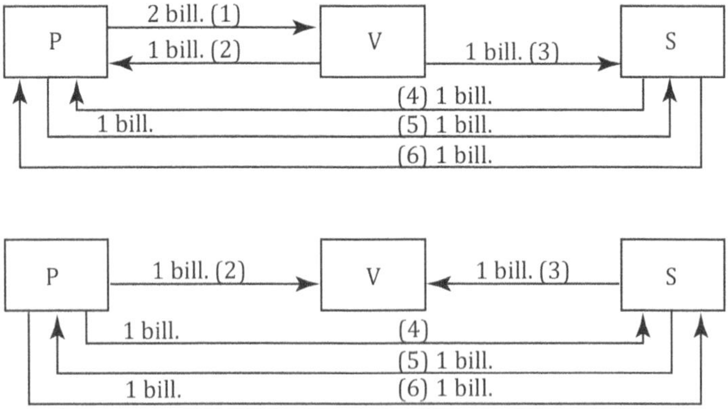

Fig. 5.1 Quesnay's scheme of the circular flow.[3] Source: Author's presentation of Marx's reconstruction of Quesnay's scheme of the circular flow. Note: bill. = billion livres; P = productive class; V = productive class/proprietary class; S = sterile class

incomplete. It included only the relations between goods and money in the processes of exchange and consumption within simple reproduction, but the relations established in other mutually related and interrelated activities, such as production and distribution, in both simple and extended reproduction were absent.

Marx improved significantly on Quesnay's work on the circular flow of economic activity. Based on the labour theory of value, Marx in the second volume of *Capital* presented the circular flow using reproduction or metamorphoses of capital as the basis of the circular flow of the total economic activity. In this work, Marx identified three forms of capital—money capital, production capital, and commodity capital—and, in turn, three connected circular flows or metamorphoses dependent on the form of capital to or from which this capital is transformed. These three types of metamorphoses are mutually interrelated and comprise the circular flow of capital.

Money Capital: The circular flow of money capital, representing the unity of trade and production, was represented with the scheme:

$$N - R \ldots P \ldots R' - N,$$

where

N – money capital
R – commodity capital
P – production capital
R' – realised commodity capital

Or in more detail:

$$\left(N - R < \begin{matrix} R_s \\ S_p \end{matrix} \right) \ldots P \ldots R'(R + r) - N'(N + n),$$

where

R_S – labour power
S_p – means of production (instruments and objects of labour)

r – surplus value (in commodity form, realised in the production process)
N' – realised money capital
n – surplus value (in money form, realised after the product is sold)

Marx distinguished the three stages. The first and third stages were within trade, and the second in production. The first stage, $\left(N - R < \begin{matrix} R_s \\ S_p \end{matrix}\right)$, represented the conversion of money capital into productive capital by using advanced money capital to purchase the corresponding means of production on the commodity and labour markets. The allocation of the elements of production realised in this manner resulted in an interruption of the circulation of the capital advanced in the form of money. Since the capital has acquired a bodily form in which it could not continue to circulate, it must enter into productive consumption. The allocation and transformation of money capital into labour power and means of production reflected qualitative, but also quantitative, relations.

The second stage of the circular flow of capital was the productive stage. This stage began with transformation of money capital into productive capital or, in other words, the interruption of the circulation of capital in trade phase and the continuation of its circular flow during production phase. In the first stage, the engaged labour power utilised the instruments of labour and applied it on the object of labour to produce commodities. Under the labour theory and following the difference between the required and actually invested labour, this created surplus value, the excess value of the product beyond the value of the elements used for its creation. The product was no longer a mere commodity, but a commodity augmented by the surplus value. Its value was the sum of the value of the productive capital utilised for its production, plus the surplus value produced by the productive capital. At the end of this stage of production, the commodity became a commodity capital as a functional form of existence of capital value that has already produced surplus value, which stemmed directly from the process of production: $R' = R + r$.

In the third stage of this circular flow ($R' - N'$), money capital, which has been transformed into productive capital and later into commodity capital, was transformed back into a money capital form by selling the commodities on the market, totalling capital value plus surplus value. This showed the capital value has produced surplus value, which is the purpose and result of the entire process of the circular flow of money capital.

The result presented a circular flow of money capital, since the opening and recurring point at its entire process is industrial capital in its money form. As such, N' (realised money capital), as a result of the transformation R'–N' in the final stage of $N - R$, was in the same form as it was at the beginning of the circular flow. As soon as it completes the three stages, it can begin new circular flow as expanded or accumulated money capital. Here Marx emphasised the distinction between the transformation, $N - R$, and the converter, $r - n$. The value capital, started its first circulation process as $N(N - R)$ and returned through the calculation $R - N$ to the same form. In this form, the second R has a different usable form, but the same value as the first R. In the second transformation the surplus value, r, created in the production process, entered the commodity market for the first time in the form of a commodity.

This completed one cycle or turnover in the circular flow of money capital. Its successive recurrences resulted in initiating the corresponding circular flows of productive and commodity capital:

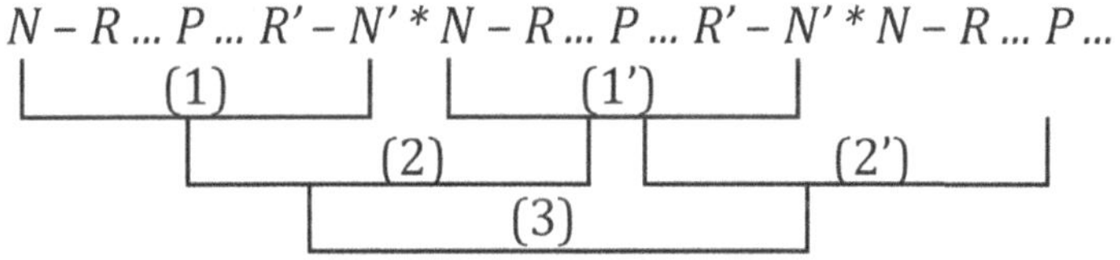

Marx concluded that for the individual capital that enters the circular flow for the first time, the first stage, N–R, represented the preparation stage of that circular flow. It was not a presumption of the production process, but was necessitated by that process.

Productive capital: Marx expressed the circular flow of productive capital as:

$$P \ldots R' - N' \cdot N - R < \begin{matrix} R_S \\ S_p \end{matrix} \ldots P(P')$$

This represented a periodic renewal of its function as a process of value expansion. The circular flow could unfold as a simple or extended reproduction, depending on the purpose of the money form of the surplus value, *n*, realised in the third stage of the circular flow of money capital and on its subsequent circular flow as capital or as means of payment used for the purchase of commodities for personal consumption.

If the owner of capital did not transform the surplus value, *n*, into the advance capital for the next circular flow round, but used it for personal consumption, the process can be shown by the following scheme:

$$R'\begin{pmatrix}R\\+\\r\end{pmatrix}\begin{matrix}-\\-N'\\-\end{matrix}\begin{pmatrix}N\\+\\n\end{pmatrix}\begin{matrix}-R<\begin{matrix}R_S\\S_p\end{matrix}\\ \\-r\end{matrix}$$

The result was simple reproduction, in which the circular flow of productive capital took the form:

$$P\ldots R'\begin{pmatrix}R\\+\\r\end{pmatrix}\begin{matrix}-\\ \\-\end{matrix}\begin{pmatrix}N\\+\\n\end{pmatrix}\begin{matrix}-R<\begin{matrix}R_S\\S_p\end{matrix}\ldots P\\ \\-r\end{matrix}$$

If, however, surplus value were transformed into advance capital, the means of production expanded, enabling extended reproduction. Here, the circular flow of productive capital took the form:

$$P\ldots R'-N'-R'<\begin{matrix}R_S\\S_p\end{matrix}\ldots P'$$

In both cases, the circular flow of productive capital deviated from the circular flow of money capital in two significant areas. First, while in the initial form, $N \ldots N'$, the process of production, a function of *P*, interrupted the flow of money capital and acted only as a mediator between its two phases $N - R$ and $R' - N'$. In this case, the circulation process (P … P') merely mediated between the productive capital that opens and closes the circular flow.

Second, the entire circulation unfolded in a contrary direction when compared to the circular flow of money capital. While the circulation of money capital proceeded according to the formula, $N - R - N$ ($N - R$. $R - N$), the circular flow of productive capital proceeded by the formula, $R - N - R$ ($R - N$. $N - R$), which is the form of the pure circulation of commodities.

Commodity capital: The circular flow of commodity capital was given as:

$$R' - N' - R \ldots P \ldots R'$$

Or in a more detailed form:

$$R' \begin{pmatrix} R- \\ -N' \\ r- \end{pmatrix} \begin{matrix} N - R < {}^{R_S}_{S_p} \ldots P \ldots R' \\ \\ n - r \end{matrix}$$

The diagram was analogous to Quesnay's scheme of the circular flow of economic activity. It differed from the corresponding circular flows of money and productive capital primarily in that it was only the commodity capital circular flow that opened with augmented capital value, not solely with capital value that had yet to be augmented. Therefore, the form $R' \ldots R'$ represented the consumption of the entire commodity product in the form of personal and productive consumption as the condition of the normal circular flow of capital.

This helped explain the difference between the conditions and possibilities of the circular flow of commodity capital, compared with the circular flows of money and productive capital. In the circular flow of money capital, this capital could expand, depending on the intended use of the surplus value, n, while productive capital could open a new circular flow with the same or even smaller advanced value and still represent a reproduction on an extended scale. In contrast, in the circular flow of commodity capital, this capital, in the form of commodity, was the precondition of production. Without the produced *R*, the circular flow was obstructed and must be reproduced as the second *R′* of some other industrial capital. It was the initial, transitory, and concluding point of the circular flow and a permanent condition of reproduction.

In this sense, Marx pointed out that although different forms of capital began all three circular flows in the same form to which it returned at the end of the circular flow, in the circular flow of commodity capital the transformation did not refer only to the functional form of capital, but also its magnitude.

From that he concluded that the circular flow of commodity capital had two specific features, which are central for the entire reproduction process. First, because the circular flow, $R'...R'$, presupposed the existence of other industrial capital in the form of $R=Rs+Sp$, with Sp aggregating diverse other capitals, it was not only the general form of the circular flow, but also the form of movement of the sum of the individual capitals, where movement of every individual capital intermingled with the other movements and was conditioned by them.

Second, because the starting point in $R'...R'$ was the total product or the total value, the extended reproduction could occur only if the part of the surplus value, which should already be capitalised, contained the material elements of the additional productive capital.

In summarising the main characteristics of the circular flows and metamorphoses of the three forms of capital, Marx pointed out the unity of production and circulation, within and between these processes. The repetition of the circular flow in one form reflected the circular flows in other forms, and every point of movement "appears as a point of departure, of transit, and of return."[4] The three circular flows or the forms of reproduction assumed by the three forms of capital unfolded side by side and continuously, constituting in their unity the circular flow of the entire social capital.

The unity and continuity of this circular flow depended on the two conditions:

- Unity of circular flows may exist only if each form of capital can go through the successive stages of the circular flow in sequence and pass from one functional form to another or, in other words, only if the industrial capital that united these parts existed simultaneously in the various phases and functions and described all the three circular flows at the same time point.

- Continuous metamorphoses of one form of capital into another require the sale of commodities at their value, but this should happen under unchanged circumstances. Marx wrote that "the process takes a wholly normal course only when the value relations remain constant; its course is practically normal so long as the disturbances during the repetitions of the circuit balance one another."[5]
- The opposite takes place when those prices fall. Money capital is set free. The process takes a wholly normal course only when the value relations remain constant; its course is practically normal so long as the disturbances during the repetitions of the circuit balance one another.

Notes

1. François Quesnay, *Analize ekonomske tablice* (Zagreb: Kultura, 1952) and Karl Marx, *Theories of Surplus Value*. Vol. 1 (New York: Prometheus Book, 2000).
2. The description of Karl Marx's theory in this chapter is based on *Capital*. Vol. 2, Chapters 1–4, trans. I. Lasker (Moscow: Progress Publishers, 1956).
3. To present this diagram clearly, the circular flow of money is displayed separately from the circular flow of production. With the exception of the transaction labelled (1), they both represent the components of the unique commodity-money transactions.
4. Karl Marx, *Capital. Vol. 2, The Process of Circulation of Capital*, Chapter 4, *The Three Formulas of the Circuit* (Moscow: Progress Publishers, 1956) 58.
5. Ibid., 62.

References

Marx, Karl. 1956. *Capital. Vol. 2, The Process of Circulation of Capital*. Translated by I. Lasker. Moscow: Progress Publishers.

———. 2000. *Theories of Surplus Value*. New York: Prometheus Book.

Quesnay, François. 1952. *Analize ekonomske tablice. Ekonomisti XVII i XVIII stoljeća*. Zagreb: Kultura.

6

Components and Course

To further understand the circular nature of economic activity, it is necessary to deconstruct it into its core components and examine their dynamics or course. The circular character of economic activity implies a continuous string of periodic metamorphoses of economic categories that unfolds normally only under appropriate conditions. Each metamorphosis comprises two core levels: economic event and economic process.

An economic event is the main element of economic activity for a society. It comprises or establishes relations between two or more economic subjects in any sphere and on any hierarchal level of economic activity, as well as between the economic subjects and nature. An economic event is not necessarily a homogenous, indivisible element, and, indeed, it usually contains a series of sub-events. These sub-events, independently, may not have an integral economic content, but acquire content through a functional merger with one or more other sub-events and through their integration into a specific economic event.

The set of all economic events experienced by a single economic subject comprises the economic activity of that subject.

Ž. Rohatinski, *Time and Economics*, DOI 10.1007/978-3-319-61705-3_6

Next, an economic process is a set of functionally related economic events. Altogether, three types of functional relations among economic events can be established within a process:[1]

- *Succession*: Creating cause-and-effect relationships among various economic events, in which one event is caused by the previous events and has consequences for upcoming events.
- *Vertical influence*: Interaction of similar or diverse economic events relative to the same sphere of economic activity, but with participants at different hierarchical levels.
- *Horizontal aggregation*: Bringing together similar types of economic events, which could include one subject, such as aggregation of gradual use of income for purchasing a single product or service, prompt expenditure for purchase of different commodities and service, or some combination, or multiple subjects, including a society as a whole, for instance aggregation of supply or demand of economic subjects.

Because of their interdependence, the sum of all economic processes within a society represents the total economic activity of that society.

The economic process is not a mechanical sum of economic events, but instead is dependent on a functional relationship established among them. The economic process determines the general conditions and premises resulting in the economic event, its participants, scope, and modalities. But this does not necessarily mean that a certain economic event was motivated by the reasons and objectives arising from an economic process. In the same time, individual economic events, to a greater or lesser extent, influence the general course and dynamics of the economic process.

In addition, economic processes need not be complex. Simple economic processes combine similar or different events occurring on the same sphere of economic activity. Complex economic processes, on the other hand, combine groups of economic events that have already been included in simple economic processes. As a result, complex economic processes represent a functional unity of the simple economic processes across different spheres of economic activity.

The economic course is the flow of economic events and related processes that form a society's aggregate economic activity. And from this understanding of the components of economic activity, the primary characteristics of economic course can be presented:

- The historical level of development of the productive forces and productive relations in a society determines the basic material and social frameworks, premises, general motives, and objectives of economic activity.
- The hierarchical structure of society originates from the level of development of its productive forces and relations and defines the main modalities of economic activity, the mode and intensity of the influence of social structures or individual economic subjects, tier relationships, and the actual motives and goals of the individual economic subjects.
- Moving to higher levels of economic activity changes the meaning of economic categories and, by extension, the motives and goals of the activity.
- The circular flow of economic activity implies that a rounded system of cause-and-effect relationships among the modalities of behaviour of economic subjects requires certain properties and a magnitude of economic categories on one or more levels transitioning from one circuit to the next.
- The main element of economic activity is an individual economic event, which creates an economic process when functionally related to other events, and the functional unity of economic events represents the economic activity of a society.
- Economic activity unfolds continuously, and any changes are dialectic.

By bringing these characteristics together, the overall course of economic activity can be described. To start, an economic event unites the material and social characteristics of the economic activity of a society and, indeed, of its total activity. The event's content, types of participants, magnitude of participation, and intensity and mode of manifestation are all determined by a dominant mode of production, the economic organisation of society, the motives and objectives inherent in the economic system, its position in relation to other events in the same economic process, economic and non-economic elements that are outside the reach of participants in this event, and the subjective preferences of these participants.

Since a specific economic event is surrounded by other events and belongs to an economic process or processes, the individual event also exerts an influence on its environment and the processes themselves. The intensity of this impact depends on the hierarchal level at which the event has been created and the levels of its participants.

Economic events unfold in a sequence and relate to one of the spheres of economic activity: production, distribution, exchange, or consumption. And since they are functionally aggregated and causally related, they create an economic process of a certain density or volume of events within a process, breadth or number of subjects, and range of characteristics. The flow of each process is circular and, if the aggregation level is below the macroeconomic level, this flow is also parallel with other processes. Naturally, this implies neither simultaneous starting points nor identical durations.

The functional integration of processes from different areas of economic activity within a given number of economic subjects forms the circular flow of these processes. This flow represents a uniform process of economic reproduction of interrelated subjects, and the functionally related processes of economic reproduction on the level of the entire society represent the economic activity in its circular flow.

Economic activity takes place in specific material and social conditions. Within that process, economic activity simultaneously affects these conditions, such as the development of productive forces or relations in production, and in that way it functionally integrates with the overall social activity. Since any such activity unfolds in the historical sequence, economic activity within its flow also follows this sequence as cycle circuits occur continuously.

Under these circumstances, the social awareness of the mode in which an economic activity occurs is dependent in the adequacy of the system of economic information. The same holds for awareness of actual decisions made by individual economic subjects. The system's adequacy represents the relevant set of information required to monitor the dynamics and the evolution of the economic system in relation to the position of the system in its environment and changes in its structure and function.

Such an economic information system contains individual bytes of information that represent a cognitive reflection of economic events, processes, or the overall activity of economic subjects; information flows between the sender and receiver of economic information; recollection of information on economic subjects; and information variables as a unity of information flows and memory contents of economic subjects.[2]

According to Theil (1967), the economic information system comprises diverse elements, ranging from types of information to the time perspective. For example, types of information include flows of money or the transfer of purchasing power between economic subjects, pricing information or a numerical value for information variables expressed in monetary units, and non-pricing information, such as a description of the content, characteristics, or purpose of the product, resource, technology, economic proceedings, and or some similar factor.

Another element of an information system is mode of reflection, which may be direct, describing an event or process from the subject's actual field of economic activity, or indirect, describing an event or process from the control field of economic activity. Similarly, the mode of temporal reflection of an economic event is also a core element of the system. Modes can be either single-phase (a certain economic event) or multiphase (a series of successive reflections over time as a consequence of an economic event).

The number of senders or channels for receiving the information is also important to an economic information system, as is the level of detail of the information, with greater detail describing an economic event in a more disaggregated manner.

A final core element of the economic information system is the time perspective. Reflection can occur ahead of an event, simultaneously, or subsequently.

The economic system is very complex and functions differently under different circumstances. An individual's understanding of it is guided by an information system that strives to reduce uncertainty as economic subjects make decisions. Together, these imply that the structure of the information system must also be complex.

The individual components of this structure or information categories are not autonomous and do not possess a pure manifestation, but rather they are perceived analogously to the categories whose behaviour or reproduction they reflect and constitute a complex information mechanism. By influencing its economic environment, this mechanism is an integral aspect of the economic activity and allows the synthesis between the conscious subjective behaviour of economic subjects and the objective impact of the rules under which economic activity occurs.

Notes

1. Extending the analysis of economic activity from individual economic subjects to wider associations, including a society as a whole, and modelling of this activity result in establishing global relationships between the aggregated elementary economic subjects, associations, variables, and parameters for their behaviour. In this sense, the endogenous and exogenous macroeconomic variables, y and $x\Lambda$, represent either a sum of adequate micro-economic variables of the state in the unit of time or a weighted average of micro-level variables of the economic course (Henri Theil, *Linear Aggregation of Economic Relations* [Amsterdam: North-Holland Publishing, 1974] 10–26).

    Problems occurring in aggregating certain micro-economic sizes, such as in cases of aggregation through transactors, commodities, and assets, were described by Leijonhufvud (1968).

    Algebraically, the macroeconomic variables represent only an aggregate of the same type of micro-economic variables. In contrast, in the case of parameters, complex relations are formed between and within the micro-economic and macroeconomic parameters, as well as in instances where macroeconomic parameters depend not only on the same type of micro-economic parameters, but also on all relations within the micro- and macroeconomic activity of a society.

    Parallel to these algebraic relations, there are also other quantitative-qualitative relations among the micro- and macroeconomic variables and parameters. For economic activity on the micro- and macroeconomic levels, the vertical relation of relative significance of individual variables and parameters is especially important. The lower the level of specificity of an individual economic variable or its corresponding parameter in relation to the number and breadth of economic subjects, the higher the level of this individual variable or parameter. In other words, the level or the relative significance of an individual variable and parameter within the economic activity is a function of breadth or density of economic subjects on the behaviour of which they homogeneously—uniformly, but not necessarily with the same intensity—influence.

Functional dependence is also formed between the micro- and macroeconomic objectives as the need to achieve macroeconomic objectives determines the conditions, frameworks, and the manner of achieving the micro-economic objectives. This is shown in Adam Smith's observation that community service was based on the egotism or self-interest of the engaged individuals. Later economic analysis, however, showed that endeavours to maximise the micro-economic objectives through mutual competition or conflicting objectives of individual micro-economic subjects would not necessarily lead to maximising the macroeconomic objectives because that would require participation of subjects on a higher level of the hierarchical structure of the economic system.

2. Based on János Kornai, *Anti-equilibrium* (Zagreb: Centar za kulturnu djelatnost, 1983) 58–59, 70–84, 208, 265.

References

Kornai, János. 1983. *Anti-equilibrium*. Zagreb: Centar za kulturnu djelatnost.

Leijonhufvud, Axel. 1968. *On Keynesian Economics and the Economics of J.M. Keynes*. New York: Oxford University Press.

Theil, Henri. 1967. *Economics and Information Theory*. Amsterdam: North-Holland Publishing.

———. 1974. *Linear Aggregation of Economic Relations*. Amsterdam: North-Holland Publishing.

7

Causal Relations

The historical and dialectic characteristics of economic categories are, by definition, reflected in the primary features of the events and processes in which they are realised, as well as the framework of possible events and processes unfolding within the historical mode of production and social reproduction. The result is a relationship in which economic categories influence economic events and vice versa. In other words, economic categories define the general characteristics and contents of economic events and processes, while processes define the relations among the economic categories realised within them. Mathematically, these relations can be represented as functions.

A causal relation is one of three interrelated types of relations that can be identified, the other two being structural and equilibrium relations.[1] Causal relations among economic categories are often referred to as the relations of behaviour and defined as the relation expressing the changes in the value of one or a set of dependent economic variables influenced by changes in value or structure of their related independent variables. This can be expressed as a function: $y = f(x)$. However, such a mathematical definition of the causal relation reflects only a portion of its quantifiable aspects, and the notion of causality essentially indicates a complex

Ž. Rohatinski, *Time and Economics*, DOI 10.1007/978-3-319-61705-3_7

qualitative-quantitative connection of economic and non-economic categories within the economic activity.

According to Bohm (1957), causality is inherent in the genesis and existence of an individual category and in the characteristics that define this category and influence how it behaves in its economic and wider social environment. As a result, causality is an important feature of every economic category and a necessary condition for establishing adequate relations between a specific category and the entire economic activity, in which every category has its place and role.

For instance, the category "commodity," defined as a direct unity of use value and exchange value, contains a number of cause-and-effect relations in the processes of production, exchange, and consumption. As a whole, these relations present the commodity as an economic category. By acquiring certain characteristics and purpose in the production process, the commodity has a potential use value, but not directly for its producer. Its use value manifests itself only through consumption.

A mechanism is needed to transfer the commodity from the sphere of production to that of consumption, as well as the simultaneous process of indirect transformation of this commodity into use value for its producer, whose means of consumption exist in the use value of other people's commodities. The commodity must enter the process of exchange, which enables it to exist as a circulation or exchange value. This way, according to Marx (1993), a commodity may achieve its use value only by realising itself as an exchange value, and, on the other hand, it may realise itself as an exchange value only by reaffirming itself as a use value in the course of exchange.

The definition of "commodity" as an object of creation, transfer, and use infers causal relations among production, exchange, and consumption, and the definition of this category may be unambiguously defined only within this framework. In the same way, all other economic categories infer within their definitions a set of causal relations resulting in their specificity.

Since economic categories may be realised through economic events and processes, their causality is also a basis of establishing cause-and-effect relations between the events and processes within the economic activity. From this perspective, the circular flow of economic activity may be seen as a main current of an uninterrupted series of cause-and-effect relations, which, through their character and manner of formation, determine the sequence and characteristics of the activity.

Therefore, causal relations between economic categories represent transmission mechanisms for the total of economic activity. Depending on the type of individual economic event or process, economic categories contained in an event or process can be necessary or even sufficient as causes, conditions, or both of the behaviour of other economic categories, while at the same time representing the result of the activity of their economic environment. This means the circular flow of economic activity also represents the cycle of causal relations.

Such relations between economic categories and among economic events and processes can be ambiguous, but are not deterministic. Depending on the concrete conditions in which they are formed, these relations have an ambiguous character from the perspective of both, cause and effect. According to Bohm, the ambiguity of causal relations has two aspects:

- Uni-multivocal: In different conditions, the same cause or a group of causes may, within a domain defined by the causal characteristics of the individual economic category, result in different effects or consequences.
- Multi-univocal: A combination of different causes may result in the same effect or consequence.

These characteristics of causal relations produce a stochastic character for economic activity.

Considering the characteristics and role of the causal relations, by definition, the economic categories, as elements of the certain economic system, also represent its active elements. They gain and confirm this status in actual economic practice, where being "active" implies that each individual economic category influences its environment and other economic categories from the domain of its causal relations and vice versa and that each active element has at least one entry and at least one exit relation with its environment, where the exit relation is a function of the entry relation.

The exit-entry function is usually called a transformation operator and expresses a mathematical rule according to which the entry relation vector is transformed into the exit relation vector of this element. In practice, this transformation may be quantitative or quantitative-qualitative, depending on whether its character is only value-related or has a material component or whether the individual active element in the transformation chain transfers onto the next element only its economic value—full, partial, or augmented—or a material substance is also transferred.

All types of transformations of economic categories signal an effort to overcome the difficulties and resistances that occur within economic activity. These obstacles must be overcome to ensure the flow of this activity and to confirm that the elements in an economic system are active and, in turn, exist as economic categories. They vary in their nature and intensity. The primary problems they address are related to the level of harmonisation of relations among the objectives of economic activity and the scarce resources required for success, differences within and among the partial interests of economic subjects, and myriad other problems that are always present in the processes of production, exchange, and consumption.

At the stage of purchasing the production factors, the difficulties primarily include sustaining the required amounts of money capital, the chance of a market shortage of certain factors of production, and the prices and terms of engagement for these factors. At the production stage, the difficulties include inefficient use of all of the advanced amounts of the production factors because of organisational and other difficulties, the potential loss of internal and external prerequisites for completing the production process, such as rain for agriculture, and the quality of completed products. The main difficulty faced in product realisation is the financial solvency of the subjects that are demanding products of the specific type, quality, and price.

Since these obstacles must be overcome for an efficient economic cycle, the permanent transformation of economic categories and the accompanying realisation and practical confirmation of their causal properties are preconditions for the existence of these categories and economic activity.

In hierarchical economic systems and the economic structure of a society, numerous vertically and horizontally interconnected elements are needed to sustain different levels of integration and influence wielded by individual economic subjects, different social-economic relations that form the backdrop for economic activity, the modalities of the listed transformations of economic categories, and the content and value of the transformation operators. From the sum of their activity, these elements should enable an uninterrupted transformation of economic categories.

This uninterrupted transformation is achieved through market mechanisms and the conscious activity of economic subjects, whether in the same or different hierarchical levels. In total, this activity should also ensure quantitative values of transformation operators are established that will:

- At least minimally satisfy the economic interests and needs of all economic subjects, which through their activity form economic events and processes that are necessary conditions for their participation in this part of the total economic activity.
- Create a tendency toward averaging the value of the transformation operators existing in different phases and within the transition points among individual phases of activity of the same type of economic subjects, essentially allocating and reallocating relative to the movement of the individual economic categories.

This is possible only when each individual transformation operator implicitly contains either direct or indirect integrated quantitative relations among all functionally related categories relevant for the circular flow of economic activity on some hierarchical level. This also creates a circular functional relation among transformation operators in the system. As a result, this mechanism becomes the qualitative modulator of the circular flow of reproduction of economic categories and establishes a unity between the qualitative and quantitative characteristics of economic categories because only the fulfilment of the certain qualitative and quantitative requirements that result from the economic activity enables the individual element category to exist as an active element in the system (see Appendix A).

On the other hand, the complexities in transformations of economic categories generally prevent transformations of the whole or the total value of economic categories from taking place promptly or even in the same circuit of reproduction. Such transformations witness a time lag, and, by extension, the causal relations between economic categories also imply a time lag between the initial causes and their consequences. These time lags are attributed to technological, psychological, and institutional factors, which are related to the origination and content of difficulties in transformation of economic categories and often occur at specific stages of economic activity:

- Technological factors occur primarily in the stage of production.
- Psychological factors arise from the inertia in behaviour of the economic subjects and their expectations for the subsequent condition of the economic activity and its flow and usually occur within exchange and consumption.
- The institutional factors reflect the social conventions and norms of the economic behaviour.

Because of the circular flow of the economic activity, the sequential time lags in the transformations of these categories create a rounded progression of the total time lags within an individual cycle of the flow of reproduction of economic categories and among many cycles. This is usually presented by a simplified scheme of time lags in the causal relations among income, demand, and end of production or the realisation of the commodities produced.[2] The archetypes within this time lag are:

- The Robertsonian lag between receiving income and using this income for consumption or the realisation of effective demand. This leads to unintended savings.
- The Lundberg lag between changes in the demand and response in output. This leads to unintended investment.
- An output-income lag between money inflows based on the realisation of produced commodities and payments in the form of income for the production factors.

The successive repetition of circular flows of reproduction of economic categories and the phase shifts in transformations of part of the value of these categories form direct causal relations among economic categories realised in various circular flows of reproduction. Economic categories in one circular flow are realised when they contribute part of their value through the phase shift to the values realised in these categories in the subsequent circular flows of reproduction. Because of their causal relations, these categories influence the formation and value of other economic categories in the subsequent circular flows. At the same time, they are themselves formed under the partial influence of economic categories realised in earlier circular flows of reproduction.

Notes

1. For more details, see Henri Theil, *Introduction to Econometrics* (Englewood Cliffs, NJ: Prentice-Hall, 1978).
2. For more details, see Allen (1968).

References

Allen, Roy George Douglas. 1968. *Macro-Economic Theory*. New York: Macmillan.

Bohm, David. 1957. *Causality and Chance in Modern Physics*. Philadelphia: University of Pennsylvania Press.

Marx, Karl. 1993. *A Contribution to the Critique of Political Economy*. Moscow: Progress Publishers. On-line version: Marx.org.

Theil, Henri. 1978. *Introduction to Econometrics*. Englewood Cliffs, NJ: Prentice-Hall.

8

Structural Relations

Unlike causal relations between economic categories, which determine the course of economic events, economic processes, and economic activity as a whole, structural relations form among functionally related categories within one or more economic events at the same stage of the circular flow of economic activity. These relations stem from economic, natural-technical, or use-character diversity of elements-subsets, economic categories of the lower level of functional aggregation, which create a main set, economic categories of the higher level of functional aggregation. Mathematically, structural relations represent a unity of subsets or a cross section of the main set.

However, analogous to that of causal relations, the economic significance of structural relations exceeds its mathematical definition. Since these relations are historically determined, each economic category-subset of a specific main set, as well as the main set itself, is causally related to other economic categories. Further, these relations as a whole determine the circular flows of the metamorphoses of these categories and their reproduction as a segment of the total economic activity.

Because of the heterogeneity of categories-subsets, the causal relations among them and other economic categories differ, implying that circular

Ž. Rohatinski, *Time and Economics*, DOI 10.1007/978-3-319-61705-3_8

flows of reproduction of these categories are different as well. As such, the structural relations between economic categories are established only in a stage of the circular flow of an economic category, the main set, in which the circular flows of economic categories-subsets are functionally integrated in a unique economic process, technical-technological process, or some combination. In other words, at this stage, the segments or phases of circular flows of economic categories-subsets are subsets of a unique process, which is perceived as the main set.

In this manner, the structural relations between economic categories represent a factor of qualitative-quantitative homogenisation or diversification of the partial economic flows and of the relevant economic categories. They represent a bridge that functionally links different economic flows into a unique economic activity of a society.

Marx (1956) illustrated the significance and role of the structural relations between economic categories on a structurally divided example of the circular flow of money capital as:

$$N - R < \begin{matrix} R_s \\ S_p \end{matrix} \ldots P \ldots R'\left(R + r\right) - N'\left(N + n\right)$$

Four structurally complex categories presented in Marx's *Capital* are:

- *Production factors*: Labour power, instruments of labour, and objects of labour.
- *Productive capital*: Variable capital, in value terms the value of labour power utilised in production and in material terms the human labour deployed by the capital value; constant capital, the value of labour instruments utilised in production; and circulation capital, the value of objects of labour utilised in production.
- *Commodity value*: Transferred value, the value of utilised objects of labour and the depreciation of utilised instruments of labour and wages, and surplus value, the excess value of a product above the total value of elements entering into its production, essentially the value of labour over wages.

- *Commodity use-character*: Means of production, commodities as they enter the productive consumption; articles of utility, commodities for personal consumption of the class of labourers; and articles of luxury, commodities for personal consumption of the capitalist class.

In the first phase of the circuit of money capital, *N-R*, the organiser of the economic activity purchases certain amounts of the instruments of labour, objects of labour, and labour power. By doing this, the homogenous monetary category, money, is transformed into a heterogeneous category in terms of value and material form, the factors of production. Structural relations are established within and among the subsets (labour power, labour instruments, objects of labour) and the main set (production factors), depending on different economic, nature-related, technical, and other characteristics of subsets and their main functions. These characteristics include availability, substitutability, mobility, quality, and price of the individual factors, technical-technological characteristics of production, expected surplus, and relations among these factors to other production processes of goods with equivalent, similar, or substitutable purpose.

These relations may be categorised in two interrelated subgroups, which can be loosely called value relations and technical relations. Value relations are formed between the value of the individual factor of production and the value of the total production factors. They indicate the presence of each factor in the subsequent production process, providing the existing process with the characteristics of an intensive or extensive process through labour power, instruments of labour, and objects of labour.

Technical relations are formed among the utilised production factors by displaying their technical characteristics:

- A relation between the utilised instruments of labour and labour power, representing the natural indicator of the technical equipment of labour.
- A relation between the objects of labour that may be processed in one flow of the production process and the utilised labour instrument, indicating the technical capacity.

- A relation between the objects of labour that will be processed in one flow of the production process and the utilised labour power, indicating the productivity of labour.

Significantly, before entering the production process and productive consumption directly, the purchased production factors represent merely a productive potential or a precondition for initiating the production process. This means that structural relations among them, which are real from the aspect of exchange, are only potential in the context of production.

In the second phase of the circular flow of money capital, the production process, *P*, the materialised labour is transformed into capital by adding labour power to the means of production, and production factors transform into the productive capital. During metamorphoses of the main set and its subsets, their structure is also transformed. These structural relations are now no longer simply value-related and technical, as seen with production factors, but primarily functional in character. This results from the different functions of individual subsets of the production capital creating value and use value of a commodity.

By entering into production, the objects of labour are transformed into circulation capital. From the material aspect, they insert their own substance into a new product. From the value aspect they lose their entire use value and their exchange value, and they transfer their entire advanced value to the value of the commodity.

By definition, instruments of labour do not enter the material contents of the commodity or circulate in their usable form, but only their value circulates. From this, according to Marx, they transfer only the part of their advanced use and exchange value (depreciation) to the commodity value, completing the transformation into the constant capital, where part of the value remains fixed in the instruments of labour. The transfers occur inversely to the number of turnovers in which these means function. The mechanism points to characteristics common to the circulation capital and constant capital: they both maintain the same value in the production process and in creating commodity value, they represent constant capital. In this sense, they differ only in the mode of value circulation and its transfer to the commodity value.

By introducing labour power into the production process, labour power's entire advanced capital value is transferred by means of production to the product and so describes both metamorphoses in the process of exchange. In contrast to the means of production, actual labour creates a new value, which comprises the equivalent of the capital value advanced into the labour power or wages and the surplus value resulting from the excess labour above the labour socially required for the reproduction of labour power or the value of labour power. The portion of capital converted into labour power changes its value in the production process and, from the aspect of creating commodity value, represents a variable capital.

Drawing from these two functional characteristics, this capital contains two functional substructures. These substructures exist because different components of productive capital behave differently as product factors and value factors.

The first substructure arises from the difference in material forms of productive capital. In the process of creating an individual product one of these forms is consumed entirely, and the other only gradually. As a result, the first substructure is the result of the varying manners in which different components of the productive capital transfer their value to the product. There are two subsets of this structure:

- Fixed capital or transformed labour instruments, which, once advanced and entered into the production process, remain in material form during their period of technical functioning. As a result, there is no need for their replacement or material reproduction in subsequent circuits of money capital. According to Marx, this means that during their useful life these instruments contribute to the production of commodities intended for circulation, without withdrawing the elements of their own renewal from this circulation. During the period in which their function is detained by means of production, the capital value advanced into constant capital goes through the circular flow of its forms only through its value, and this occurs partially and gradually. A part of its value remains in circulation and is transformed back into its natural form.[1] Repeated transformation of money into the natural form of the means of production happens only at the end of its functioning period.
- Circulating capital, transformed labour power and objects of labour, whose elements are entirely consumed in the production process. At the end of each circuit of money capital, these elements are reproduced or renewed in nature, implying that the value of the circulating capital

is advanced only for the time required to finish the product. This value is completely invested into the product and, by its sale, returns from the trade as a whole and may be advanced again.

Comparison of structures of productive capital and production factors confirms that structure of productive capital is directly linked to relations in the technical structure of production factors, such as capacity and available technical equipment. As such, this structure represents their transformed expression within production. Also, the relations in the structure of the production factors formed during exchange, which were considered to be potential from the aspect of the production, now become real or actual.

Here, the relation between the circulating and constant capital formed in the first circuit of money capital indicates the size of the capital value. Along with the unchanged relative relations of value, size, and technical-technological characteristics of the production and its factors, the relation must be advanced in each subsequent circuit to maintain the unchanged level of capacity. This shows the amount of the augmented capital value that should be re-invested at the end of each circuit into the subsequent circuit, in which labour instruments function in unchanged volume and quality.

The second substructure of productive capital results from the different characteristics of productive capital elements in forming product value. Productive capital is divided into two components:

- *Constant capital*: The portion of capital that does not change its value in the production process when transformed into means of production.
- *Variable capital*: The part of capital that is transformed into labour power and changes its value in the production process. It reproduces its own equivalent and the surplus value, which could also vary. This portion of the capital is constantly changing from the constant value into a variable value.

Within such structure of the productive value, two relations are formed. The first is between the constant and variable capital or, as Marx called it, the "organic composition of capital," and the second is between the surplus labour and necessary labour or the "rate of exploitation."

According to Marx's analysis, these relations are crucial throughout reproduction in capitalism. In their totality, they determine the profit rate and, for the total advanced money capital, the level of profit, the key driver of production and main criterion for investment of moncy capital.

These relations are determined ex ante from the distinct combination of deployed production factors, their efficiency and relative prices, and the historical price of labour power. They also influence, in turn, the scope and structure of deployment of production factors, and by extension the crucial features of the production process and total reproduction.

At the end of the production phase and after the impact of productive capital, a commodity or a group of commodities, $R' (R+r)$, is created with its use-character and value structure.

The use structure of commodity contains elements intended for reproduction of the means required for new production, such as instruments and objects of labour, and commodities intended for consumption, such as articles of utility and luxury. Marx's analysis found that this structure is the basis of the structural division of production of the entire economy into sectors, provided that the commodities produced in the circuit of individual capital have a certain level of homogeneity. It is also directly causally related to the structure of demand, and, thus, to the totality of the value relations formed in all phases of the circular flow of capital. These relations provide the use structure of commodity with a special significance for the total reproduction: They form in a single circuit of the total capital circulation, and any change in these relations in the subsequent circuits reflects the product specialisation of the economy, which is closely related to the level of development of their productive forces and the available natural resources and to the specific objectives and motives of the production process, as well as the total of the corresponding reproduction.

The use structure of commodities created as a result of the activity of productive capital of specific structural characteristics is not simply a reflection of needs within the structure of consumption.[2] Essentially, it represents a complex set of values and material relations established in all phases of the capital circuit, the conditions in which the circuit unfolds, and the motives and goals of its subjects.[3]

The value structure of commodities stems directly from the structural relations formed among the elements of productive capital, which are based on their different characteristics in forming value. The value of constant capital and the value of variable capital, corresponding with the value of necessary labour, are transformed into the transferred value, while the surplus labour, which is materialised in the actual commodity and created by the variable capital, represents the surplus value.

The relation between the surplus value and the transferred value established in a single circuit of capital, as determined by the organic composition of capital, exploitation rate, and depreciation rate, represents the potential profit rate of this circuit. In total, it also represents the potential profit rate of the total individual capital advanced for all circuits in which the instruments of labour were advanced at the onset of the first circuit.[4] The potential character of that profit rate results from the potential difference between the commodity value and the effective price that the commodity realises on the market, that is, the price that results in an effective profit.

Finally, in the third phase of the circuit of money capital—exchange, *R'–N'*—the commodity acquires the use value for its owner or consumer through its sale in the market, while its value transforms into the form of money. Depending on the existing structure of the commodity value, this money has two monetary forms:

- Monetary equivalent to utilised production factors.
- Additional money capital or profit as a transformed form of the surplus value.

The advanced capital repeatedly acquires a money form through which it can begin a new reproduction process. Depending on its use-character, it could be divided in two subsets, resulting from either simple or extended character of reproduction.

The first subset contains advances for simple reproduction of the means of production and labour power, money form of the depreciation of labour instruments accumulated in the reserve fund, and funds intended to sustain the capitalist consumption. Advances and money form are alimented from the monetary equivalent of utilised production factors.

The second subset represents accumulation or difference between the size of the additional money capital and funds intended for capitalist consumption and serves as an advance for utilising the additional production factors.

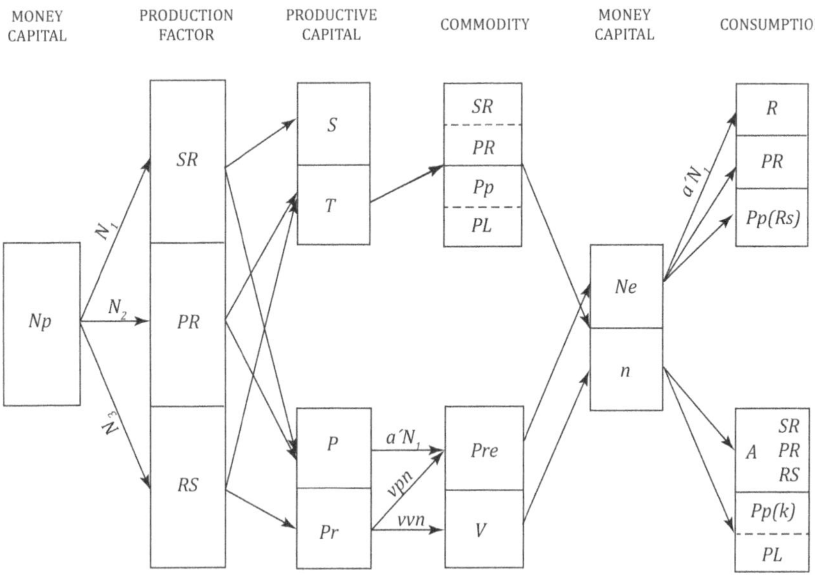

A = accumulation
a' = rate of depreciation
N_p = advanced money capital
N_e = monetary equivalent to utilised production factors ($a'\ N1 + N2 + N3$)
n = additional money capital (profit)
P = constant capital
P_r = variable capital
P_{re} = transferred value
P_R = objects of labour
P_p = articles of utility
P_L = articles of luxury
R = reserve fund (depreciation)
RS = labour power
SR = labour instruments
S = fixed capital
T = circulation capital
V = surplus value
vpn = value of necessary labour
vvn = value of surplus labour
s = rate of accumulation (savings)

Fig. 8.1 Structural relations in the reproduction process. Source: Author's presentation

The totality of structural relations established in the circuit of money capital may be presented in a scheme the flows from money capital to consumption (Fig. 8.1).

The scheme for the structural relations in the circuit of money capital shows that the circuit has three rounded circular flows of reproduction of its main subsets, as well as the circuit of profit. Their main forms are:

- Circuit of money capital advanced into instruments of labour.

$$N_1 - S_R < \begin{matrix} S \\ P \end{matrix} a'N_1 P_{re} - N_e\left(= a'N_1\right) - R\left(\sum a'N_1 = N_1\right) - S_R - \ldots$$

- Circuit of money capital advanced into objects of labour.

$$N_2 - P_R < \begin{matrix} T - commodity \\ P - P_{re} \end{matrix} > N_e\left(= N_2\right) - P_R - \ldots$$

- Circuit of money capital advanced into labour power.

$$N_3 - R_S \left\langle \begin{matrix} T - commodity \\ P_r \left\langle \begin{matrix} P_{re} - N_e(=N_3) - R_S - \cdots \\ V - n \end{matrix} \right. \end{matrix} \right.$$

- Circuit of profit.

$$N_p \left\langle \begin{matrix} N_1, N_2 \\ N_3 - R_s - P_r - V - n \left\langle \begin{matrix} A(=\Delta N_p) \left\langle \begin{matrix} \Delta(N_1, N_2) \\ \Delta N_3 - R_S - \cdots \end{matrix} \right. \\ articles\ of\ luxury \end{matrix} \right. \end{matrix} \right.$$

Comparing the circuits of these subsets of money capital with the circuit of money capital as the main set shows that they completely fulfil the circuit of the main set, implying that these circuits are the subsets of the circuit of money capital perceived as the main set. This suggests a direct interdependency of the circuits of the subsets of money capital and the circuit of money capital. The circuit of the subset may be realised only if the circuit of the main set is realised and, in turn, the circuit of the main set may be realised only by realising the circuits of all its subsets. At the same time, the structural relations among the elements and subsets of the main set, among those in the main set, and among the circuits of subsets and the circuit of the main set dominantly determine the character, properties, and motives of the historical form of economic activity.

These conclusions are derived from the example of the circuit of money capital and from the example of the structure and structural relations between the subsets or categories that are heterogeneous in terms of content and function within the key economic category, the main set. Still, they are not absolute and cannot be applied to the circuits of all structurally complex categories or to their structures by taking into account all of their features. However, they are sufficiently indicative to point out the significance and function of structural relations for economic activity, representing its connective tissue in all important aspects.

From here, it is possible to define more closely the characteristics of the circuit of the total advanced money capital from the perspective of the interrelations among causal and structural relations established in this circuit between the functionally related economic categories. Different qualitative and quantitative characteristics of categories or subsets in which the advanced money capital is transformed into the main set also condition the different partial courses of their reproduction. These reproduction courses differ from one another across several features:

- Duration, depending on the natural-technical features of the certain categories, especially circular and fixed productive capital.
- Contents and characteristics of the production process of the individual purpose-type products.
- Phase shifts.
- The level of gradualism in realisation and transformation of these categories.

However, in terms of money capital reproduction, different economic categories represent only partially transformed forms. Because of this, the circular flow of reproduction of their values expressed in the form of money may be reduced to a uniform shape in which all differences among them are essentially the quantitative aspect of their duration. In this form, the average duration of a single circuit of the total advanced money capital is an average duration of the circular flows of reproduction of its components transformed into different economic categories or subsets.

In this way, the individual circuit of reproduction of the total advanced money capital represents a weighted average of circuits of its components. And further, the circuit's duration is not fixed, but varies when there are different individual capitals, as well as when it is in different stages of reproduction or development of the same capital. This variance depends on a number of factors, including organic composition of capital, capacity, technical equipment of labour, productivity of labour, and economic life of labour instruments.

In addition, the duration of an individual circuit of the total advanced money capital is statistically derived and does not necessarily match the actual duration of the circuit of reproduction of any of the partially transformed parts of the initially advanced total money capital. It contains several complete circuits of reproduction of the parts of the initially advanced total money capital. This capital is transformed in the circulating productive capital and comprises only a corresponding part of a single circuit of reproduction of the part of the initially advanced total money capital that is transformed into fixed productive capital.

Moving onward, with a simplifying assumption that the circulating productive capital is a homogeneous category, an average duration of a single circuit of reproduction of the total advanced money capital, DT_{ϕ}, is determined by:

$$N\frac{1}{DT_{\varnothing}} = N_S\frac{1}{DT_S} + N_T\frac{1}{DT_T} \qquad \left(N_S + N_T = N\right)$$

This can also be expressed graphically (Fig. 8.2).

Under this definition, a single circuit of the total advanced money capital contains more than one circuit of reproduction of its part which is transformed into circulating production capital and so establishing the causal and structural relations among economic categories within a single circuit of the total advanced money capital should primarily be viewed as an aggregated expression of individual relations among the economic categories provided in each individual segment of the circuit of total money capital and in which the complete circuit of reproduction of the circulating productive capital is realised.

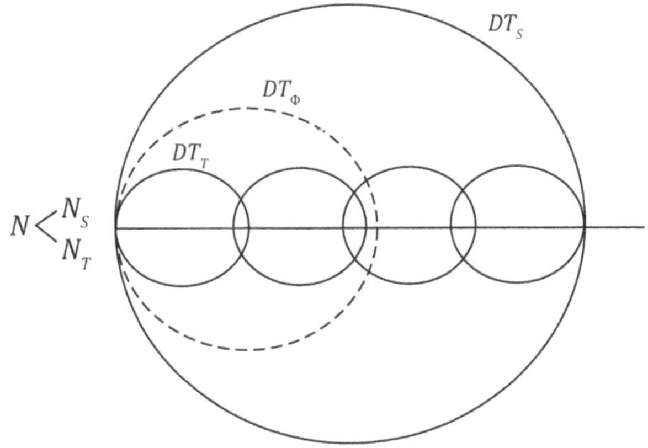

Fig. 8.2 Average duration of a single circuit of reproduction. Source: Author's presentation

This mechanism does not change the values of transformation operators and the structural relations among the functionally related economic categories within the circuit of the total advanced money capital when compared with their values formed within the part of the circuit, which corresponds to the duration of the circular flow of reproduction of the part of the money capital transformed into the circulating capital. In that way, the consistency of Marx's approach to the analysis of these relations is preserved.[5]

The paramount advantage of such an approach is that it moves from the partial analysis of the actual circuits of the individual parts of the total advanced money capital to the aggregate analysis of their joint course in the certain segment or period of the reproduction process in which the partial circuits are causally and structurally homogenised as a uniform circuit. The duration, content limits, and values of the economic categories in this segment or period are not defined externally or arbitrary, but arise functionally from the totality of interrelations of economic categories existing within the certain reproduction process. As a result, they are not just a statistical indicator, but very important to both the conditions and the manner of reproduction of the individual money capital, as well as the relations among the reproduction of the two or more individual money capitals.

Notes

1. This occurs because the part of the value of labour instruments transferred to commodities, which is transferred back into the form of money upon its realisation on the market, is at the onset of the following circuit of money capital (if labour instruments were not fully utilised) not advanced for the purpose of renewal of these instruments, but, according to Marx, accumulated into reserve funds of money.
2. Marx believed the size and structure of demand were not determined exogenous to the structure of production and production relations in a society. In his view, demand does not represent the "social demand" that would *a priori* define the size and the purpose-related structure of production, but instead class relations in a society and, therefore, the ratio of surplus value to wages and relations in the distribution of surplus value essentially determine social demand. Marx acknowledges that demand for commodities appeared "at first sight" to be at a level set by society to sate its own desires, but then pointed out that this demand is very elastic and subject to the whims of factors such as changes in prices and wages. Beyond this, capitalists also influence consumer demand by, for instance, producing more cotton-based products as price or cotton drops to capture more surplus value (Karl Marx, *Capital.* Vol. 3, *The Process of Capitalist Production as a Whole* [New York: International Publishers, On-line version: Marx.org, 1996] 138).
3. Relations in the purpose-related structure of production included in Marx's schemes of extended reproduction have been discussed by many later theorists, who looked particularly at the causes of economic crises and whether the underlying "purpose" of capitalist production was production or consumption. The prevailing view leaned towards labour instruments of the first sector as the driving force behind production rather than the production of goods for consumption.

   However, technological advances, along with changes in the rate of accumulation and societal structure (the development of hierarchical systems), have overtaken this perspective. For example, in 1976 Branko Horvat said the growth rate of the total production, r, was a function of the share of production of the first sector in the total production, $s = P1/P$; the share of material expenses and expenses of the replacement of labour instruments in the value of production, $u = U/P$; and the capital coefficient, k, and its change over time, k'. From this, the growth rate of the total production would be: $r = (s - u)/k - k'/k$. This result allowed for various combinations of production growth rates in the first and second sector, whether production were growing, falling, or stagnant. This combination depended on the features of the "technical parameters" and preferences of an organiser or moderator of the production process (Branko Horvat, *Odabrane teme iz ekonomske analize* (*Selected Topics from Economic Analysis*) [Zagreb: Faculty of Foreign Trade, 1976]).

4. Marx uses this indicator with an assumption of the total advanced capital turned over in one circuit.
5. Viewed in this context, the rate of depreciation, a', and relative size of capacity, k, for an average circular flow of reproduction of the total advanced capital, may be defined as:

$$R = N_S : DT_{\varnothing},\ \textit{and}\ \text{therefore}\ a' = R : N_S$$

$$k = \sum_{1}^{DT_{\varnothing}:DT_{T(PR)}} N_{T(PR)} : N_S$$

References

Horvat, Branko. 1976. *Odabrane teme iz ekonomske analize (Selected Topics from Economic Analysis)*. Zagreb: Faculty of Foreign Trade.

Marx, Karl. 1956. *Capital. Vol. 2, The Process of Circulation of Capita*. Moscow: Progress Publishers.

———. 1996. *Capital. Vol. 3, The Process of Capitalist Production as a Whole*. New York: International Publishers. On-line version: Marx.org.

9

Equilibrium Relations

Causal and structural relations between economic categories are the main transmission and connection mechanisms of economic activity. These mechanisms functionally integrate all economic events, processes, and activities of subjects in economic reproduction. But to enable the integration and ensure the normal reproduction of economic categories realised through these events and processes, these mechanisms must result with a value- and purpose-related harmonisation of interrelated economic categories or, in other words, with partial equilibriums within the economic system. Since normal reproduction of economic categories or subsets is a condition of the normal reproduction of the aggregated economic category, the main set, the relations of partial equilibriums must be harmonised among these and ensure the overall equilibrium of the system.

Princeton economist Fritz Machlup,[1] among others, has defined the equilibrium relation as a state within a certain point, phase, or even the entire circular flow of economic activity at which selected interrelated categories are harmonised to prevent a tendency towards

Ž. Rohatinski, *Time and Economics*, DOI 10.1007/978-3-319-61705-3_9

changes in a system they compose.[2] Such a definition leads to three crucial conclusions:

- The overall equilibrium of the economic system does not require a balance among all interrelated individual economic categories, but only among those that are important for the functioning of that system.
- To establish an equilibrium, all categories in the system must be in a state of relative rest. Moreover, the state of the relative rest of each individual category must be compatible with the state of rest of all other categories in the system.
- The state of rest results only from balancing the internal relations in the system, while the influence of the external factors remains fixed.

In line with French economist Léon Walras' premises, critical prerequisites for an overall equilibrium of an economic system include a perfect knowledge of the market situation possessed by all economic subjects, perfect competition, direct exchange of goods and services, an expectation of stable prices, and closeness of the system.[3] With this understanding, the general equilibrium of an economic system represents a specific constellation of interrelated economic categories. This raises a crucial question: What is the relationship between this constellation, the state of equilibrium in economic systems, and other possible constellations of economic categories formed within the circular flow of economic activity?

Two dominant views have emerged within economic theory to address this. In the first, the equilibrium view, balance is a natural or normal state of an economic system and it periodically or continuously occurs during economic activity. This balance is brought about by an automatic balancer among economic categories within a given economic system, a reflection of classical and neoclassical theory,[4] or corrective activity within the hierarchical organisation of economic activity where higher-lever structures influence lower-level structures of a society.

Depending on the location of the state of equilibrium within economic activity and its extensibility, analysis of the equilibrium view can be approached from two perspectives. The first approach perceives the general equilibrium of an economic system as the state in the certain point or phase of the circular flow of economic activity that is periodically renewed. Here, the circular flow of economic activity represents the transfer of economic categories from one equilibrium state to another as their natural relation. This transfer or the system's need to move from one state of equilibrium to another can last indefinitely if the conditions remain unchanged and is the result of disturbances in the initial equilibrium of the system brought by changes in the external conditions.

Any changes, however, require the formation of a new equilibrium or constellation of interrelated categories within the system. The resulting general equilibrium is referred to as the static equilibrium and the method used to analyse its formation and the behaviour of the economic system as the balance is established is known as comparative statics.[5]

The second analytical approach considers the general equilibrium as a normal constellation of interdependent categories in motion, implying that the system is continuously in a dynamic equilibrium. This means that in the circular flow of economic activity there are no disturbances of the equilibrium relations among economic categories established at the starting point of this course. This view of the general equilibrium of the economic system as a normal or at least realistically possible constellation of interdependent categories in motion inspired numerous models of equilibrium growth, including those devised by Roy F. Harrod and Evsey Domar, Nicholas Kaldor, Richard Kahn, Joan Robinson, neo-classicists, and others. It has also drawn serious criticism centred on the initial basis of these models and the concept of general equilibrium, either in the state of rest or in movement of economic categories.[6]

Contrary to the equilibrium view, a second dominant view in economic literature suggests that the course of economic activity alone leads to disturbances in the initial equilibrium between the economic categories. From this view, the disequilibrium view, the economic system continuously exists in different levels of imbalance and the circular flow of economic activity represents a chain of disequilibria.[7]

Viewed from the perspective of the equilibrium-disequilibrium of the system, the course continuously goes through four phases:[8]

- *Initial situation*: System in equilibrium.
- *Action*: Disequilibrium or change in the system.

- *Reaction*: System adjustment.
- *Closing situation*: New equilibrium.

In the disequilibrium view, static conditions are only seen in the first and fourth phases, with the two intermediate phases representing the movement of economic categories within the system. The static or equilibrium state cannot actually exist within the continuous flow of economic activity, which is constantly in disequilibrium. This conclusion implies that static states only exist hypothetically and represent the relations between the economic categories that the system is trying to establish, but never fully achieves.

From this perspective, the hypothetical equilibrium state of the system is only static from a relative perspective and represents what Samuelson called "simply degenerate special cases" of dynamical systems.[9] The purely hypothetical existence of equilibrium in the circular flow of economic activity is static only from the aspect of the transforming interim phases, which mark the system's departure from and approach towards the equilibrium state, but not from the aspect of the motion of economic categories and functioning of the economic system.

The functioning of the economic system in the environment characterised by disequilibrium relations between economic categories in motion is usually presented by the example of a dynamic multiplier in a multiplier-accelerator model, which is important in explaining the economic cycles.

While this exposition, of course, does not attempt to settle the debate between the equilibrium and disequilibrium views, it raises two significant questions in considering the time dimension of economic theory:

- Is it possible that while the economic system exists in or strives towards a general equilibrium for there to be a disequilibrium in one of the markets within the system?
- How do different characteristics and roles of various production factors in the reproduction of an economic system influence the formation or non-formation of the general equilibrium of the economic system or, in other words, is the existence of the system in equilibrium state or a periodic tendency towards such a state necessary for normal reproduction of economic categories and the system as a whole?

The first of these questions focuses on John Maynard Keynes, while the second may be examined using Marx's schemes of reproduction.

In Marx's analysis of the circular flow of economic activity, the general equilibrium of an economic system represented a regulated set of relations within and among the value-structure and use-structure of production, which resulted with equilibrium between supply and demand of goods. His conclusion was derived from the schemes of simple and expanded reproduction, created under specific premises:[10]

- The value structure of the total number of commodities produced, W, as well as of each individual commodity, comprises the constant capital utilised in production, P; variable capital, P_r; and surplus value, V.

$$W = P + P_r + V$$

- The purpose structure of produced commodities comprises the means of production and means of consumption. Following that, the total production is, in terms of use, divided into production of the means of production in the first sector (I) and production of the means of consumption in the second sector (II).

$$W = W(I) + W(II), \text{ whereby}$$
$$W(I) = P(I) + P_r(I) + V(I)$$
$$W(II) = P(II) + P_r(II) + V(II)$$

- The production value of the first sector is twice that of the second sector.

$$W(I) = 2W(II)$$

- The rate of exploitation ($se = V/RS$), is constant and equal in both sectors or 100 per cent.

- The organic composition of capital, $w = P/P_r$, is constant, and in case of simple reproduction equals 4 in both sectors, while in case of expanded reproduction equals 4 in the first sector and 2 in the second sector.
- In relation to the constancy of organic composition of capital, in expanded reproduction the use of accumulation or new investment has the same structure as the capital in the previous circuit.

$$A_{(P)} : A_{(\mathrm{Pr})} = P : P_r \; or \; \Delta P : \Delta P_r = P : P_r$$

- There is no transfer of accumulation between sectors, meaning that the accumulation achieved in one sector is invested in the same sector.
- The system is closed, and there is no exchange between this system and other systems.

Following these assumptions, the condition of general equilibrium of the system is that the production value of the first sector equals the total demand for the means of production in both sectors, while the production value of the second sector equals the total demand for the means of consumption. Also, representing exchange between the sectors, the demand for the means of consumption in the first sector equals the demand for the means of production in the second sector.

In the simple character of reproduction, the general conditions of equilibrium of economic system in Marx's analysis are:

$$P_r(I) + V(I) = P(II) \tag{9.1}$$

$$P(I) + P(II) = W(I) \tag{9.2}$$

$$P_r(I) + V(I) + P_r(II) + V(II) = W(II) \tag{9.3}$$

In expanded or recession reproduction, they are in the first circuit:

$$P_r(II) + \Delta P_r + V(I) = P(II) + \Delta P(II) \tag{1}$$

$$P(I)+\Delta P(I)+P(II)+\Delta P(II)=W(I) \quad (2)$$

$$P_r(I)+\Delta P_r(I)+V(I)+P_r(II)+\Delta P_r(II)+V(II)=W(II) \quad (3)$$

From the point of view of location and extensibility, the general equilibrium of an economic system is in equilibrium in the point or exchange phase at the end of the current circuit of capital and at the beginning of the subsequent one. This equilibrium is static, but because of dynamic schemes of reproduction it may be transformed into the equilibrium of economic system in motion, a dynamic equilibrium.

However, in both the static and dynamic interpretation the general equilibrium of an economic system presented in the schemes of reproduction may be established at the point of transformation from the current into the following circuit of money capital. This occurs only provided that the value of utilised instruments of labour, which in nature must be replaced at the beginning of the subsequent circuit of capital, is equal to the value of the calculated and paid depreciation from the previous circuit. Otherwise, at the point of transfer between the circuits of capital there will be neither equilibrium between the supply of labour instruments and their demand nor between the supply and demand for the means of production:

$$P(I)+P(II)\neq W(I)$$

Consequently, exchange between sectors could not be evened out and a disequilibrium between supply and demand for the means of consumption is created.

The sectors of reproduction could also be viewed as two newly employed individual capitals. In this case, while maintaining the hypothesis that such double-purpose production system is closed, a *conditio* sine qua non of establishing the general equilibrium is to ensure that labour instruments used in the first circuit are fully utilised in the same circuit, in other words their depreciation rate is 100 per cent. This condition is in direct conflict with the very definition of labour instruments as fixed capital.

Faced with this problem, Marx reasoned that some additional assumptions must be considered to reach a solution:

- The system is viewed from a macroeconomic aspect, which means that reproduction sectors are not uniform, but represent sets of individual capitals that independently pass the circuits of their reproduction. These circuits vary in terms of their beginning, components, and duration.
- Reproduction does not start ab ovo, but each individual circuit has a prehistory, either in earlier circuits of the same capital or in the circuits of other individual capitals.
- Under these assumptions, different capitals placed in various branches of industry have different ages.
- The unit of observation is not an individual circuit of any partial capital, but the average circuit of the total capital, which Marx set at a duration of one year.
- There is a relatively rigid relation between the portion of fixed capital that must be renewed in kind within the observation unit and the portion of fixed capital that continues to function in its unchanged form.

Although these are logical assumptions, they contain several weaknesses that should be highlighted:

- Even with these assumptions, it is not possible to explain how the initial ab ovo equilibrium of an economic system is established.
- The presence of a relatively fixed relation between the portion of fixed capital to be renewed in kind and the portion that continues to function in its unchanged form could be a coincidence rather than a rule.
- Additional assumptions tend to undermine the premise that an economic system is closed, which is core to the definition of a general equilibrium of an economic system.
- The approach leads to the conclusion that the inability of a system to create a general equilibrium in the observation unit triggers an automatic reproduction crisis. This would inevitably result in serious disorders in its functioning, regardless of whether these refer to over-

production of labour instruments or to an accumulation of stocks and search for liquid capital reserves.

More troublesome, even if such circumstances were accepted, a standard definition of the notion of general equilibrium remains unsettled.

A more realistic approach would be to offer a definition that is more flexible, even if it adds complexity. Namely, the difficulties in relations between the value of current production of labour instruments and the effective demand for them, which occur at the point of transfer between the circuits of capital, also reflect complications in at-par realisation of other dedicated product groups when capacity and level of use are unchanged. This would a priori restrict the possibility of establishing a general equilibrium in terms of equal supply from the current production and demand in any individual point or phase of the circular flow of reproduction. In the same time, equilibrium conditions defined in the schemes of reproduction, although they can be fulfilled at the point of transfer between the two circuits of capital only by coincidence, represent the framework in which the economic system functions normally. For this reason, the imbalances between supply and demand occurring at the transfer points between the circuits of capital have a periodic tendency to completely or predominantly neutralise each other.

The mechanism could be illustrated by the examples of simple and expanded reproduction based on the extreme assumption that the production sectors are composed of two compact individual capitals and total supply and demand for commodities in the system is formed within the reproduction of these two capitals. Using Marx's original numerical assumptions, these examples show that in simple reproduction, the cumulative equilibrium of a system is formed at the end of the circuits of capital when the utilised labour instruments are replaced in both sectors.

At the same time, they also show that this does not occur in expanded reproduction. In expanded capital reproduction, the general equilibrium conditions are not fulfilled at the point of transfer between any two circuits of capital, which mirrors simple reproduction, or in the sum of the circuits.

The imbalance between supply—production plus stock—and effective demand for commodities of the first sector is constant and inherent to the functioning of the system in the given initial circumstances. In this sense, the relative size and structure of this disequilibrium display two basic characteristics:

- The relative size of the cumulative disequilibrium[11] between supply and demand of the first and second sectors for commodities of the first sector grows hyperbolically. This means the disequilibrium gradually stabilises with a tendency to become fixed when the value of hyperbola gets significantly close to the value of its asymptote. Parallel to the movement of the relative size of the cumulative disequilibrium, the relative sizes of the current disequilibria between production and demand for commodities of the first sector decrease in each individual circuit of capital, with a tendency to stabilise when the relative size of the cumulative disequilibrium becomes stable as well.
- The relative size of the cumulative disequilibrium between supply and demand of the first sector for commodities of the same sector remains constant during the entire capital reproduction, while the relative size of the cumulative disequilibrium between supply and demand of the second sector for commodities of the first sector grows following a hyperbolic curve. Changes in the relative size of the total cumulative disequilibrium result exclusively from the changes in the relative size of the cumulative disequilibrium between the supply of commodities of the first sector and the demand for these commodities by the second sector.

The characteristics of disequilibrium between supply and demand for commodities of the first sector and thus of the general disequilibrium in the system are directly connected to the relations among the dynamics of production growth, its value, and the structure of its allocation, which are established between the first and second sector in the course of capital reproduction. As a result, this disequilibrium is not partial, but stems from the totality of causal and structural relations existing among economic categories in economic system.

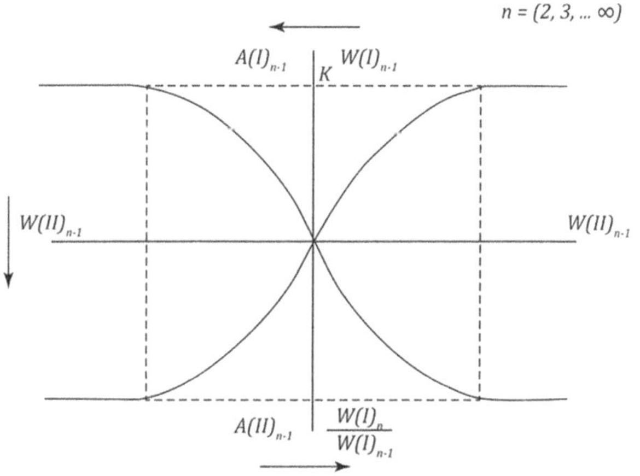

Fig. 9.1 Conditions of steady growth in a two-sector system. Source: Author's presentation of steady growth in a two-sector system based on the model results

On this basis, two interconnected sets of relations emerge between economic categories in the functioning of the presented economic system:

- A set through which the initial relation between the production value of the first and second sector in the circuit of capital influences the growth rate of production of the second sector in the subsequent circuit of capital (Fig. 9.1).
- The influence of changes in relations between the growth rates of production of the first and second sector on the changes in the share of unrealised commodities of the first sector in the production value of the first sector and, by extension, on the relative magnitude of disequilibrium in the system (Fig. 9.2).

With other elements fixed, the growth rate of production of the second sector depends on the relation between production value of the first and second sector, making this relationship critical to the totality of expanded capital reproduction. Reproduction stabilises with the fixed relative mag-

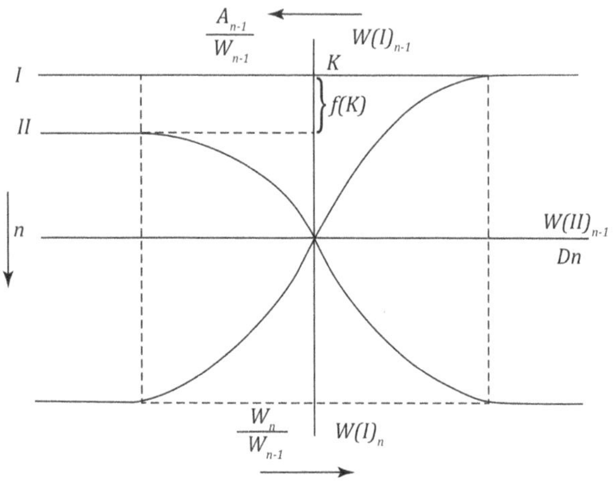

Fig. 9.2 Relative magnitude of disequilibrium in a two-sector system. Source: Author's presentation of relative magnitude of disequilibrium in a two-sector system based on the model results

nitude of the system disequilibrium seen only in the circuit of capital that follows after the circuit in which the relation between production value of the first and second sector has reached a fixed magnitude, a point at which the rates of subsequent growth in production of the first and second sector equalise.

As such, even with the extreme assumption of a system composed of only two individual capitals, economic categories may be re-produced in this system and economic activity may unfold normally, even under the conditions creating a permanent state of disequilibrium, as well as the equilibrium conditions as defined by Marx, or in other words when the equilibrium among economic categories cannot be established in either of the individual circuits of capital or in either of the cumulative circuits of capital, as seen in expanded reproduction (see Appendix B).

* * *

This analysis of equilibrium relations leads to a series of conclusions:

- The forms and modalities of establishing the general equilibrium of a system depend greatly on its structure and the mode of economic activity in the system.
- Equilibrium may be established at an individual point, phase, or period of economic activity only in the special case. In principle, equilibrium is established as a whole and for the whole of the course of economic activity and does not refer specifically to any partial segments—event, phase, or period—of this course.
- All relevant economic categories in a system, representing its economic potential—whether material, human, and financial—participate in establishing the equilibrium. Along with permanently active elements, these include elements that are temporarily passive in the certain phases and periods of economic activity, for example because of time lags.
- In these conditions, Machlup's definition of general equilibrium of an economic system is too restrictive to include all elements important for normal reproduction of economic categories and activity within the system.
- A more flexible image of the general equilibrium of an economic system may be defined to replace such a constellation of selected interdependent variables arranged in a way that prevents the tendency to change. This definition creates a set of relations among the relevant economic categories in the course of economic activity that enables the normal reproduction of these categories. The notion of normality depends on the structure of a system, the manner in which economic activity unfolds, and the economic and technical-technological characteristics of individual categories.
- The system does not oscillate around some hypothetic or fixed state of equilibrium at a point, phase, or period of the economic activity that is uniformly, quantitatively, and qualitatively defined for the entire course of this activity. Instead, the set of equilibrium relations that ensure the normal reproduction of economic categories in the system is created or not created in the course of economic activity.

- If the set of equilibrium relations among economic categories is not created in the course of economic activity, then, taking into account the initial assumption of the closed system and the given economic potential, the normal reproduction of economic categories requires a change in the transformation operators and structural relations between these categories.
- Since the system exists in permanent motion because of the causal relations among economic categories, the equilibrium relations are dynamic.

These conclusions imply that the equilibrium relations essentially represent a framework for forming the causal and structural relations among economic categories, within which the reproduction of these categories takes its normal course, as well as the economic activity of a society. In this context, while the causal relations are formed among the economic categories that usually exist in different, successive, phases of reproduction and the structural relations among these categories are formed in the same phase of reproduction, the equilibrium relations are formed simultaneously among economic categories within a single phase and among different phases of reproduction. As a result, they represent a link among all economic categories in the system and regulate the functioning of this system.

Notes

1. Fritz Machlup, *Equilibrium and Disequilibrium: International Payments, Debts, and Gold* (New York: New York University Press, 1975).
2. A broader definition of equilibrium relations was given by Joseph A. Schumpeter. In his view, equilibrium represents a set of relations between economic categories, the variables in a system, that determine a series of values of the variables without a tendency to change under the exclusive influence of the facts, which are per se included in these processes (Joseph Schumpeter, *History of Economic Analysis* [New York: Oxford University Press, 1954]).
3. See, for instance, Nicholas Kaldor, *Collected Economic Essays: Essays on Value and Distribution* (London: Gerald Duckworth and Co. Ltd., 1960).

4. The conceptual basis of this approach is Jean-Baptiste Say's *Law of Markets* (*Loi des Debouches*) (New York: Augustus M. Kelley, 1971), which argues that the supply of products on the market automatically ensures the formation of the purchase funds to promote the demand for these products. Therefore, an increase in production not only leads to an increase in supply, but also to an increase in demand. See also Karl William Kapp and Lore L. Kapp, *History of Economic Thought: A Book of Readings* (New York: Barnes & Noble, 1949).
   This is mathematically expressed by a modified Keynes–Lange equation of identity:

$$\sum_{i=1}^{n-1} P_i D_i = \sum_{i=1}^{n-1} P_i S_i$$

   where:
   S_i = supply of the product i (quantity),
   D_i = demand of the product i (quantity),
   P_i = current price of the product i,
   $n-1$ = number of product
   Schumpeter proposed that if money itself is taken as n-number of product the identity equation takes the form:

$$D_n = S_n$$

   Joseph Schumpeter, *History of Economic Analysis* (New York: Oxford University Press, 1954).
   Further analysis of this model leads to the conclusion that the activity of the market mechanism under perfect competition automatically ensures the equilibrium of supply and demand on the market of commodities and production factors at full employment of the production factors. This implies the laisser-faire view of the economic reproduction in a society and makes a policy of streamlining by subjects on the higher decision-making level redundant and even adverse if it interferes with the free functioning of the invisible hand of the market.
5. Among others, Paul A. Samuelson, *Foundation of Economic Analysis* (Boston: Harvard University Press, 1971); Michael D. Intriligator, *Econometric Models, Techniques and Applications* (Amsterdam: North-

Holland, 1978); and Alpha C. Chiang, *Fundamental Methods of Mathematical Economics* (New York: McGraw-Hill, 1974).

6. One of the critiques of the theory of general equilibrium in an economic system was given by János Kornai in *Anti-Equilibrium*. He synthesized 12 groups of extreme assumptions on which individual theories of general equilibrium are implicitly based:

   - Static character of analysis and stationary character of the functioning of the economic system.
   - Immutability of the set of organisations in the system.
   - Economic system consisting exclusively of two types of organisations: producers and consumers.
   - Immutability of the set of products in the system.
   - Simultaneity of activities of producers and consumers without any time period between production, sale, and consumption and, thus, without the ability of this period to influence the prices.
   - Convex production set, for example, no indivisible products and resources, relations between the product input and output may be described by using continuous and differentiable functions, no increase in yield, while there is no increase of the marginal substitution rate between the production factors.
   - Maximisation of profits as the criterion for selection of the production alternatives.
   - Convex set of possible consumption, where the consumer has a preference-based arrangement and maximises its function of utility which is concave.
   - Constant sets of production and consumption, including the preference-based arrangements.
   - Prices as the only flow of information between organisations in the economic system.
   - Anonymity of market relations in terms of absence of any individual relations between producers and consumers.
   - Lack of uncertainty.

   János Kornai, *Anti-equilibrium* (Zagreb: Centar za kulturnu djelatnost, 1983).

7. Albert O. Hirschman, *The Strategy of Economic Development* (New Haven: Yale University Press, 1958) 62–70.

8. Fritz Machlup, *Equilibrium and Disequilibrium: International Payments, Debts, and Gold* (New York: New York University Press, 1975) 114.
9. Samuelson, *Foundation of Economic Analysis.*
10. Karl Marx, *Capital. Vol. 2, The Process of Circulation of Capital*, trans. I. Lasker (Moscow: Progress Publishers, 1956).
11. The cumulative disequilibrium can be seen as the share of the value of disequilibrium in the cumulative value of production of the first sector for the circuits of capital in which the initially advanced instruments of labour are not exchanged in kind:

$$\sum_{i=1}^{n} \frac{D_i}{W(I)}$$

References

Chiang, Alpha C. 1974. *Fundamental Methods of Mathematical Economics.* New York: McGraw-Hill.

Hirschman, Albert O. 1958. *The Strategy of Economic Development.* New Haven: Yale University Press.

Intriligator, Michael D. 1978. *Econometric Models, Techniques and Applications.* Amsterdam: North-Holland.

Kaldor, Nicholas. 1960. *Collected Economic Essays: Essays on Value and Distribution.* London: Gerald Duckworth and Co. Ltd..

Kapp, Karl William, and Lore L. Kapp. 1949. *History of Economic Thought: A Book of Readings.* New York: Barnes & Noble.

Kornai, János. 1983. *Anti-equilibrium.* Zagreb: Centar za kulturnu djelatnost.

Machlup, Fritz. 1975. *Equilibrium and Disequilibrium: International Payments, Debts, and Gold.* New York: New York University Press.

Marx, Karl. 1956. *Capital. Vol. 2, The Process of Circulation of Capital.* Trans. I. Lasker. Moscow: Progress Publishers.

Samuelson, Paul A. 1971. *Foundation of Economic Analysis.* Boston: Harvard University Press.

Say, Jean-Baptiste. 1971. *A Treatise on Political Economy.* New York: Augustus M. Kelley.

Schumpeter, Joseph A. 1954. *History of Economic Analysis.* New York: Oxford University Press.

10

Modelling Economic Activity

Among economic categories, causal, structural, and equilibrium relations together constitute a network with channels through which the reproduction of the individual categories and of the system takes place. The character of some of these channels is relatively fixed and remains unchanged through a number of circuits of reproduction, which could be determined by, for instance, the level of development of the production forces, social and economic relations, the economic organisation of a society, and characteristics of economic categories. Another group of these channels is variable and established afresh in different modalities in each individual circular flow of reproduction of economic categories depending on the actual conditions in which the specific circuit takes place. The successive repetition of the circuits of reproduction of economic categories within the economic activity of an individual subject forms the functional sequences of relations among the categories created within the economic activity of that subject.

The elements of these sequences may be grouped in sets according to the criteria of selection of economic categories among which these relations are formed or the phase of the economic activity in which they appear. Since they were formed under different conditions during circuits of reproduction, the individual relations within the set of one type of

Ž. Rohatinski, *Time and Economics*, DOI 10.1007/978-3-319-61705-3_10

relations differ both qualitatively and quantitatively to a greater or lesser extent. However, these relations also share features, expressed either fully or as a tendency in each individual circuit of reproduction.

These features substantially determine the entire set of the same type of relations and represent a rule that governs how individual relations are established among economic categories in certain phases of the circuits and economic activity of the given subject. Further, the rules that govern how total activity of this subject takes place are also functionally connected. Autonomous decisions of an individual subject, depending on its hierarchical position, influence these rules only partially and with restricted intensity. These rules largely result from interactions of all economic subjects, with which this subject forms indirect and direct relations through economic events and processes in which it jointly participates.

The parallelism and interdependence among the circuits of reproduction of economic categories and activities of economic subjects on the same and different hierarchical levels, integrated in the economic activity of the entire society through functional connections among partial events, processes, and activities, lead to general rules of forming relations among the aggregate economic categories. Thereby, the general rules of economic activities at the level of society are formed as well.

Analogous to the general relations among economic categories, the economic rules are divided into causal, co-existence, and functional.

In terms of the genesis of economic rules, method of formation, characteristics of economic categories, and relations among them, these rules have common characteristics:

- At their core, economic rules are the set of common characteristics present in relations among economic categories, which in practice form in different modalities in each individual circuit of reproduction of these categories and also within the events and processes in which these categories are realised. Further, economic rules represent an average or typical node around which the individual relations fluctuate. Following that, these rules are stochastic.
- In addition, the historical character of economic categories determines a historical character of economic rules, so they are not the same in all

social formations or in all phases of development of these formations. Their form and content are formed in the context of the concrete historical conditions in which economic activities took place in a certain society and change according to the changes of those conditions.

- These characteristics open economic rules to dialectic analysis. By existing in the conditions in which economic activity takes place and because they change based on changes in these conditions, the rules also reflect the dialectics of the economic activity.
- The economic activity of each individual economic subject is directed by specific motives and interests. At a given level of development of production forces and social relations, an individual's economic activity confronts, through a series of partial events and processes, the activities of other economic subjects taking part in these events and processes. The motives and interests of these subjects may be complementary or contrary. As a result, the content, characteristics, and modalities of the course of certain events and processes derive from the activity of economic subjects participating in them. Within these events and processes, certain economic categories are realised and relations are formed among them.

  Because these relations are the result of the activity of economic subjects within as individual event and process, the individual relations among disaggregated economic categories, formed within the partial events and processes, do not entirely depend on the subjective preferences of the participants in these events and processes. In addition, the regular relations among aggregated economic categories at the level of society also do not fully depend on these preferences. This provides economic rules with an objective character in terms of their relative independence not only from the preferences of economic subjects, but also from the awareness these subjects have of their own activity.

The set of all economic categories and their mutual relations in the course of economic activity at the certain level could be seen as a mechanism of that activity. Further, a regulated set of economic categories relevant to the activities taking place at this level and connected among themselves by what is socially perceived as regular is a model of this activity. As a result, the economic model represents a functional unity of the socially perceived general rules of economic activities taking place at the certain level.

By definition, this model comprises:

- Relevant economic categories aggregated at a specific level that are either independent or dependent variables of the model depending on their roles in causal relations. In the circular flow of reproduction of economic categories the attributes of independent and dependent have only relative or conditional meaning.
- Regular relations among these categories. These are the functional relations among individual economic categories, as conditioned by the character of these categories, a concrete modality of economic activity, and, since the model is a conceptual creation, the theoretical conception of the modality of this activity, and the parameters expressing the form and intensity of the quantitatively measurable functional relations between individual economic categories.

From the perspective of the evolution of economic activity described by the model, the listed elements or components of every economic model are, apart from being different in content, also different according to their stability. In this sense, economic categories are the most stable elements, since they exist in the whole of the certain historical type of economic activity. The least stable or most flexible are the parameters, which, as average values, vary depending on specific conditions in which the economic activity takes place. As a consequence, the historical course of economic activity of an economic subject and of a society may be described either by one model with flexible elements or by a number of models with fixed elements, where the change of elements in the mode or the change of the model indicates a phase in the evolution of the main modalities of economic activity of this subject and a society.

In this context, the main modalities of the circuits of capital reproduction presented by this hypothetic analysis can be illustrated in a scheme representing a structurally divided Marx's model with its elements aggregated at the level of the society (Table 10.1).

This scheme may be compressed using an aggregate model of economic activity of a society with the elements:

Table 10.1 Structurally divided scheme of money capital reproduction

Money capital	Transformation operator (3):(1)	Production factor	Transformation operator (5):(3)	Productive capital	Transformation operator (7):(5)	Commodity (value)		Transformation operator (9):(7)	Realised commodity and stock
(1)	(2)	(3)	(4)	(5)	(6)	(7)		(8)	(9)
	$\frac{1}{1+k+\frac{a'+k}{w}}$	SR	a'	$P(SR) = a'SR$	1	$W(SR) = a'SR$		$\frac{a'w}{(w+1+se)(a'+k)}$	$D = a'SR$
						+			
N_o	$\frac{k}{1+k+\frac{c'+k}{w}}$	$PR = kSR$	1	$P(PR) = kSR$	1	$W(PR) = kSR$	$W = (a'+k + \frac{(1+se)(a'+k)}{w}SR$	$\frac{kw+a'k}{(w+1+se)(a'+k)}$	$N = \frac{kw+a'+k}{w}SR$
						+			
	$\frac{a'+k}{w\left(1+k+\frac{a'+k}{w}\right)}$	$RS = \frac{a'+k}{w}SR$	1	$P_r = \frac{a'+k}{w}SR$	$(1 + se)$	$W(PS + V) = (1 + se)\frac{a'+k}{w}SR$		$\frac{se}{w+1+se}$	$n' = \frac{se(a'+k)}{w}SR$

(*continued*)

Table 10.1 (continued)

Money capital	Transformation operator (11):(9)	Utilised production factor and capitalist consumption
(9a)	(10)	(11)
N	$\frac{kw}{kw+a'+k}$	$PR = kSR$
	$\frac{a'+k}{kw+a'+k}$	$RS = \frac{a'+k}{w}SR$
	$\frac{s}{\left(1+k+\frac{a'+k}{w}\right)}$	$\Delta SR = \frac{s(a'+k)se}{w\left(1+k+\frac{a'+k}{w}\right)}SR$
n'		
	$\frac{ks}{1+k+\frac{a'+k}{w}}$	$\Delta PR = \frac{ks(a'+k)se}{w\left(1+k+\frac{a+k}{w}\right)}SR$
	$\frac{a+k}{w}\frac{s}{1+k+\frac{a'+k}{w}}$	$\Delta RS = \frac{s(a'+k)^2 se}{w^2\left(1+k+\frac{a'+k}{w}\right)}SR$
	$(1-s)$	$L = \frac{(1-s)se(a'+k)}{w}SR$
N'	1	$D-(RS+\Delta RS+L)$

Source: Author's presentation

(1)	$W = (a', k, w, se)(SR, PR, RS)$	Causal relations
(2)	$V = se\ RS$	
(3)	$A = sV$	
(4)	$I = A$	
(5)	$\Delta SR, \Delta PR, \Delta RS = (a', k, w)I$	
(6)	$L = (1 - s)V$	
(7)	$P + P_r + V = W$	Structural relations
(8)	$V = A + L$	
(9)	$w = \dfrac{P}{P_r}$	
(10)	$k = \dfrac{PR}{SR}$	
(11)	$r = \dfrac{V}{P + P_r}$	
(12)	$W - D = SR + \Delta SR + PR + \Delta PR + RS + \Delta RS + L$	Equilibrium relations
(13)	$D - (RS + \Delta RS + L) = N'$	

where

$P = a'SR + PR$
$P_r = RS$
D = according to (49) from Appendix B

(See Appendix C.)

By encompassing the general rules in relations among relevant economic categories, economic models represent a manifestation of the achieved level of awareness of economic subjects about the main modalities of their activity. Models explain these modalities, but they also provide an analytical basis for the conscious activities of economic subjects on different hierarchical levels aimed at directing the courses of reproduction towards the realisation of their preferences.[1]

* * *

These views on the characteristics, contents, and modalities of economic activity lead to a series of conclusions:

- Economic activity is a complex, organised, and multilevel activity of individual subjects and a society aimed at conducting and directing a number of material and social processes to ensure the basic material needs and other prerequisites for their existence and development.
- The main entity of economic activity is an individual economic event, which forms the relation among two or more economic subjects or between these subjects and nature. Divisible and structurally heterogeneous in terms of technical, legal, or sociological perspectives, this event is unique and homogenous in terms of economics.
- A functionally rounded set of economic events represents a process of the certain level of complexity, density, and modality This set embodies certain content and character and belongs to a specific sphere of economic activity—whether production, distribution, exchange, or consumption—and is interconnected by functional connections of succession, interdependence, and aggregation.
- Functional merging or concatenation of economic processes from different areas of economic activity forms the circular flow of economic reproduction of subjects participating in these events and processes. Their functional unity represents the economic activity of a society.
- When included in economic events and processes, the natural, material-technical, financial, spiritual, and other resources available to society that provide the basis for economic activity turn into corresponding economic categories or, in other words, into the categorical apparatus of this activity. This produces a complex set of economic relations among these categories that may be divided into three main groups:

    (a) Causal relations representing an essential characteristic of each economic category and the basic mechanism of making functional connections between economic events and processes by way of qualitative-quantitative transformations of economic categories into a uniform course of economic activity.
    (b) Structural relations formed among the functionally related economic categories—elements or subsets of some main set—within

the framework of a single economic event or in the same phase of the circular flow of economic activity of individual subjects. These are the basis for homogenisation or diversification of individual circular flows of reproduction of individual economic categories or economic subjects.

(c) Equilibrium relations representing a regulated set of the causal and structural relations among economic categories, required for the normal reproduction of these categories and economic activity of a society.

- Repeated repetition and the parallel unfolding of the circular flows of reproduction of economic categories within the frameworks of economic activities of individual subjects and a society result in some common qualitative and quantitative elements existing as an average or tendency within the specific phases of these flows. These elements become parts of nearly all causal, structural, and equilibrium relations of the same type formed among the relevant economic categories. Manifested on the level of the economic activity of the society, they represent the general rules of this activity, which are stochastic, historical, and objective.
- The cognised or observed set required for the unfolding of economic activity within a certain level of the key categories interconnected by the corresponding causal, structural, and equilibrium relations represents a model of this activity.

Notes

1. Jan Tinbergen, "The Use of Models: Experience and Prospect," *The American Economic Review* 71, no. 6 (December 1981): 17–21.

Reference

Tinbergen, Jan. 1981. The Use of Models: Experience and Prospect. *The American Economic Review* 71 (6, Dec.): 17–21.

Part III

Economic Activity in Time and the Character of Functional Time

Bringing together the discussion on the nature and character of time with that of the characteristics, structure, and modalities of economic activity, this analysis examines the time flow or time dimension of economic activity. In particular, it looks at the interdependence of modalities of economic activity and time flow contained in this activity and at the maturity of economic activity as dependent on specific modalities of its course in the periods of different duration.

Analysis of the time component of economic activity usually begins with the concept of time as seen by classical physics—the absolute character of time, space, and motion—and treats it as an absolute exogenous category and as a pure *milieu* in which the economic activity unfolds. From this perspective, time is not functionally incorporated into economic activity and the interaction of the two is not perceived as a condition of the functional unity of economic activity and of the time in which it takes place.

This approach is reflected in the aggregation of established values and flows of reproduction of economic categories within the selected calendar units of time, which are the main units of observation of the properties and modalities of economic activity of individual subjects and of the society. It is also echoed in the corresponding information system. In an exercise known as period analysis, collected and processed information enables economists to create an idea of the type, character, modalities of formation, and numerical values of the causal, structural, and equilibrium relations among economic categories in selected units of calendar time and also of the state of economic activity in these time units or the change of state within these units.[1]

Other authors analyse the process of economic decisions in a narrow sense and the totality of the conscious action of economic subjects in a wider sense and are more inclined to a phenomenological view of the nature and character of time. From this perspective, they emphasise that the instance of "now" or the "initial moment"—as the centre of all current experience, cognition, expectations, decisions, and actions—is always subjective and different for individual economic subjects. This "now" instance is not an isolated point of time, but rather "a short interval…, the existence of which includes its continuous motion and continuous evolution… along the calendar axis (causing the economic subjects to keep 'moving about')… towards a fresh temporal point of view."[2] English economist G. L. S. Shackle (1958) called this a "dynamic movement in time" and in his view all current transformations of one situation into another are realised through this actual translation of the present moment.

This analysis, however, looks beyond these approaches to understanding the temporal dimension of economic activity and begins with the concept of time as an objective functional category. In this light, time is a fully integrated element of economic activity and one of the most significant modulators of this activity. The analysis starts by purposing a definition of the notion of functional time and moves forward to look at its influence on the course and modality of economic activity.

Notes

1. William J. Baumol, *Economic Dynamics: An Introduction* (New York: Macmillan, 1959) 127–41; Henri Theil, *Linear Aggregation of Economic Relations* (Amsterdam, North-Holland, 1974) 39–58.
2. George Lennox Sharman Shackle, *Time in Economics* (Amsterdam, North-Holland, 1958) 14–15.

11

Functional Time

An examination of functional time as it pertains to economic activity begins by reviewing Marx's concept of the temporal component of capital reproduction in a two-sector system. Marx wrote that the movement of capital through the phase of production and the two phases of circulation or exchange takes place in a time sequence. The duration of the time spent in the phase of production constituted its production time, while the time spent in the phase of circulation constituted its circulation time. By extension, the entire time in which capital described its circular flow equalled the sum of the production time and circulation time.

This sum was a time segment that ran from the moment the capital value was advanced in the certain form until the moment of its return to the same form. The individual circuit in the life of capital represented only this repeating segment of time, that is, one period. Following Marx,[1] the circular flow of capital, seen as a periodic process rather than an individual act, was called a capital turnover. Its duration was the sum of its production time and circulation time and measured the distance between one period of the circular flow of the total capital value and the subsequent one, altogether the periodicity in the life of capital.

Ž. Rohatinski, *Time and Economics*, DOI 10.1007/978-3-319-61705-3_11

Considering different properties of the fixed and circulating capital and the gradual transfer of the value of the initially advanced money capital transfers to a commodity during the production process, the complete turnover of advanced money capital occurred only with the realisation of the commodity to which the fixed capital has transferred the remaining part of its value. Thus the duration of turnover or the turnover cycle was defined by the duration of the fixed capital advanced in the production process.

Of course, this time was neither fixed nor equal in different economic fields or in production of different groups of dedicated products, among individual capitals within these groups, or in the certain stages of development of individual capitals and the total capital in a society. Development of fixed capital extended this time, while technological development shortened it. As a result, the length of the temporal duration of the complete turnover varied depending on a number of technical, technological, and economic factors.

By dividing the reproduction time of the total advanced money capital into production time and turnover time, Marx provided a detailed analysis of the structure of each of these times. He focused on the content and properties of the part of the reproduction process taking place in this time, especially from the aspect of value creation and capital expansion.

In that sense, Marx pointed out that the time in which the production factors take part in producing a certain commodity included:

- Time of the process of labour or the time during which labour is utilised in the production process.
- Excess production time over labour time, when productive capital is in a latent state in the sphere of production and not functioning in the actual production process or it functions in the production process and is not part of the labour process.

According to Marx, this view implied that excess production time over labour time may be created on the following bases:[2]

- Periodic interruptions of the labour process and production process, for example at night if the processes of labour or production are not

continuous, which interrupts the functioning of the production factors, but not their existence at the location of production and the time spent in the sphere of production.
- Interruptions of the labour process, but not the production process, for example in agriculture, when the technological characteristics of the production process lead to interruptions in the labour process and in working time. During this time, the specific means of production are affected by physical processes without human labour, such as chemical or physiological processes. In that way, the production period exceeds the labour period, but the product is completed only upon the completion of the production period.
- The need to keep the part of the means of production, such as raw materials, ready at the location of production. The means are a condition of the production process, but have yet to enter the actual production process.

In this analysis, production time comprised the time of labour, the time in which the labour process and production process are periodically interrupted, the time in which the production process unfolds without the influence of human labour, and the time in which the means of production wait to be included in the production process.

Marx also noted that such a structure for production time resulted with differences in the modes in which the individual production factors—elements of production capital—take part in forming the commodity value. The instruments of labour—fixed capital, whose function is interrupted by periodical breaks in the production process—transferred to the commodity within the production time only the part of their value proportional to the share of production time in the length of duration of labour instruments. This means the labour instruments lost their use value and also their total value both when they function and when they do not. The objects of labour transferred their entire value to the value of the commodity during the production time because by entering the production process they lost the form required for circulation. In contrast, the labour power—whether its transformed value was expressed as time-wages or piece-wages—transferred its value to a commodity, and it created an excess value and expanded the capital only during labour time.

Under this premise, Marx made a precise functional distinction between labour time and production time from the aspect of their respective roles in the capital reproduction:

> Whatever may be the reason for the excess of production time over the labour-time … in none of these cases do the means of production function as absorbers of labour. And if they do not absorb the labour, they do not absorb surplus-labour, either. Hence there is no expansion of the value of productive capital so long as it stays in that part of its production time which exceeds the labour time … It is plain that the more the production time and labour time cover each other the greater is the productivity and self-expansion of a given productive capital in a given space of time. Hence the tendency of the capitalist production to reduce the excess of the production time over labour-time as much as possible. But, while the time of production of a certain capital may differ from its labour-time, it always comprises the latter, and this excess is itself a condition of the process of production. The time of production is, then, always that time in which a capital produces use-values and expands, hence functions as productive capital, although it includes times in which it is either latent, or produces without expanding its value.[3]

Analysing the difference between the labour time and production time in relation to the main driving force of the entire capital reproduction process or the creation of surplus value and expanding capital shows clearly that the main prerequisites for achieving this goal are formed in the time of the labour process, which represents the basic time within the reproduction process.

Viewed from the aspect of raison d'être of capital reproduction, the excess of the production time over labour time, and thus the entire production time, may be expressed as a function of the duration of labour time, although there is no—or should be no—technological relation between them. In essence, the excess of the production time over labour time may be understood as an addition to labour time through which unfinished commodities must proceed to acquire technical-technological, economic, and use properties. These properties will ensure the realisation of the main objective of capital reproduction, the prerequisites of which have been created during labour time.

However, the concept of labour time is not yet precise enough to provide a quantitative definition of the temporal length of the labour process that must impact the means of production to produce an individual commodity or a group of commodities of the same type. Marx called this a labour period and defined it as a number of connected working days required to deliver a finished commodity. A working day referred to the length of working time in which a labourer must exert labour power daily. As such, the labour period may be more simply defined as a length of working time, expressed in adequate units of calendar time such as working hours, necessary to produce a finished commodity.

Such a labour period, analogous to the structure of production time, combined with the period of excess production time over labour period constituted the period of production of a certain commodity. In the context of creating surplus value and capital expansion, this period may again be expressed as a corresponding function of the length of the labour period and in the units of labour period.

By reviewing the influence of the length of labour period on the total course of capital reproduction, Marx noted two important elements:

- Assuming the period of production is the same duration as the labour period or the period of the excess of production time over labour period has a constant duration, then shortening the duration of the labour period within the temporal duration of the entire turnover of advanced capital increased the cumulative value of production and, as a result, the surplus value. This resulted in faster and more intensive capital expansion. Consequently, a permanent desire arises to reduce the labour period by introducing more efficient instruments of labour and technology of production, improving work organisation, and increasing of specialisation, among other measures.
- Depending on the length of the labour period—assuming equal amount of total capital, a fixed relation between the fixed and circular capital, and equal duration of the working day—the amount of money capital advanced in a single circular flow of the total capital reproduction, that is the size of the circular capital advanced for production of a single commodity, also changes. This effect changes the time of one

circular flow of advanced capital, including its speed. The longer the labour period, the bigger the quantity of circular capital, which remains in the form of productive capital for a longer period that should be advanced.

Because of the varying length of the labour period in different stages of development of individual capital, the length of the labour period in a calendar period between the production of different groups of dedicated products, as well as within these groups, is also generally not equal. The length of the period depends on the technical-technological complexity of a commodity, the level of development of the production forces, technical equipment, technological characteristics of the labour process, the properties of the means of production, and other elements. At the same time, the periods of excess of production time over labour time are also not equal because they depend on the continuity or discretion of the labour process, technological and natural properties of production, and similar factors. This means that the production periods of different products, the same products produced by different producers, or the same products produced by the same producers but in different stages of development are also different.

Marx defined the time of circulation as the time during which capital stays in the sphere of circulation as commodity capital or money capital. Since both of its circulation processes include the transformation of capital from the form of commodity into the form of money, and from the form of money into the form of commodity, the time of its circulation is also composed of two parts:

- Time required for the transformation from commodity into money.
- Time required for transformation from money into commodity.

As a result, the time of capital circulation consists of the time of selling and the time of buying. From the point of the final objective and driving force of capital reproduction, time of sale is especially important since, in addition to transformation *R-N*, it includes the transformation of surplus value into surplus of money capital or, in other words, into profit, *R'-N'*.

In terms of the role and influence made by the time of circulation on the total course of capital reproduction, Marx pointed out that the time of circulation and the time of production were mutually exclusive because during the time of circulation, capital does not function as productive capital and produced neither commodities nor surplus value.

Analogous to earlier definitions of the labour period, the period of excess production time over labour period, and production period, the length of circulation time—the sum of buying time and selling time—required to buy or employ factors of production and to sell the produced commodity on the market, that is time required to realise the transformations of capital, *N-R* and *R-N, R'-N'*, represents the period of circulation of this commodity and the capital contained in it.

The duration of the period of circulation is dependent on a number of factors, including:

- Volume of the part of capital that must be kept in money form, which is determined by time of replacement of certain parts of productive capital.
- Distance from the market.
- Development of transportation infrastructure.
- Existing ratio of supply and demand on the market.
- Method of sale.
- The time in which the commodity retains its use properties.

With other conditions remaining equal, because differing lengths of circulation periods influence the difference in duration of one capital circuit, they also affect the amount of circulating capital that should be advanced in a single circuit of its reproduction.

Examining the entire process of production and realisation of individual commodity, the process evidently takes place in a period of time comprising the labour period, the period of excess production time over labour time, and the circulation period. From here, the duration of the circular flow of reproduction of capital to be advanced for production and realisation of a single commodity or a group of commodities represents a sum of the lengths of durations of these three periods and may be called a period of individual reproduction of initially advanced money capital. In other words, it may be called a period of a single circular flow of reproduction of money capital that should be advanced for production and realisation of an individual commodity.

Determining more precisely the temporal dimension of this period and its properties when looking at the interdependence of the content of economic activity taking place during this period and its temporal attributes requires a more exact definition of the concept of "duration." A starting point is Whitehead's perception of time as a sequence of epochal durations and Bloch's view that the turnover of time is an integration of "now" and "never." Using these views to analyse the properties and course of economic activity, the concept of duration in terms of economics may be defined as a functional sum of successive calendar or chronological units of time, which represent temporal attributes of functionally related economic events within the economic process.

Since the character, content, scope, and temporal-spatial modalities of the occurrence and course of these events are determined by natural, technical-technological, economic, and other properties of categories participating or being realised in these events, the overall characteristics of the process represent a specific constellation of these characteristics and the accompanying relations among economic categories. As a result of this constellation, the temporal dimension of the certain process or its duration is formed.

This duration represents a closed slice of time. It starts with the initial event followed by other functionally related events constituting a certain process, and ends with an event that interrupts this process so it can be started anew. In this slice of time, economic content is formed that will be followed by other economic processes within the circular flow of economic activity of an individual economic subject.

Economic processes of different levels of complexity comprise a certain number of economic events, which in turn comprise sub-events of different character. As a result, the structure of an economic process is heterogeneous in terms of content. Each event and its accompanying sub-events have specific temporal attributes, which determine whether they have the same or different durations, simultaneous, in temporal sequence, or in another temporal constellation. The duration of an economic process is structurally homogeneous if viewed from the aspect of temporal attributes of its corresponding economic events and their sub-events. However, all these events make sense and are economic in character only if they are functionally related in a common economic process within the total activity of individual economic subjects. This provides them with a property of unique functional totality.

Functional unity of content is necessarily connected with functional unity of the content's temporal and spatial attributes, which, according to Whitehead, compose a "pattern" of functional relations among events within a process. From this, the duration of economic processes may be defined as the time range for realisation of the entire temporal-spatial pattern of this process. This conclusion implies that, in relation to other processes and individual events, the functional unity of content, time, and space of a certain economic process results with its external compactness and homogeneity, although if viewed internally, this process is divisible and heterogenic in terms of content, time, and space.

Because the relations among economic categories are not deterministic, the constellation of these relations among the economic categories participating in a specific process is not the same in each repetition of that process. This means the duration of this process also changes, which has a reversible influence on the causal, structural, and equilibrium relations among economic categories in the process.

However, analogous to the character and determinants of forming economic rules, each individual process of the same type exhibits some common elements. The manifestation of these elements in time determines the average duration of that process. Depending on other elements, especially external ones, the effective durations of individual processes fluctuate around this average duration. Further, in relatively stable conditions for individual processes of the same type, their durations of these processes strive towards an average length of the calendar time required for the realisation of a regulated set of causal relations among the categories taking part in a given process. This results in the process duration becoming an objective category.

In that way, the duration of such periods—labour, production, or trade—within the circuit of money capital that must be advanced for production and realisation of individual commodities is defined by the calendar time of the duration of the relevant economic processes, that is, those that enable an individual product in a production process to acquire its use value and exchange value from that money capital, enabling it to be transformed into commodity capital and commodity capital back into money capital in the process of trade.

The sum of the durations of these processes represents the duration of the period of a single circular flow of the reproduction of money capital that must be advanced for the production and realisation of an individual commodity. Since such a period includes several different labour, production, and trade processes, it is not homogenous in terms of its content, time, or space. However, the functional aggregation of economic processes in the economic activity of a specific subject leads to homogenisation of the content, time, and space of the economic processes representing the economic activities of this subject in relation to the content, time, and space of the economic activity of other subjects.

In this sense, the duration of a labour period is essential for the time flow of economic activity and the related relationships between the economic categories established in a single circular flow of the money capital reproduction. Other periods or subsets within this period, the main set, might be defined as functions of the duration of a labour period:

$$\frac{PVVP}{RP} = pvvp, \quad \frac{PP}{RP} = pp$$

Under such conditions, the duration of a period of one circular flow of reproduction of money capital is defined by the function:

$$PR = (1 + pvvp + pp)RP$$

where PR = the period of circular flow of reproduction, RP = labour period, $PVVP$ = period of excess production period over labour period, and PP = period of trade.

On this basis, the turnover time of the total money capital paid in advance, with its calendar length determined by the duration of the instruments of labour, might be defined as a result of the duration of the period of a single circular flow of the money capital reproduction and a number of products that were repeatedly and discretely produced and realised in the course of this turnover. Since the duration of the circular flow of the money capital reproduction may be expressed as a function of the duration of a labour period, the turnover time of the total advanced money capital, *VO*, may also be expressed as the units of the duration of this period:

$$VO = Q\left(1 + pvvp + pp\right)RP$$

Here, *Q* represents the number of products or conditional products that were produced discretely and realised in the course of the turnover of the total advanced money capital.

Comparing this approach to measuring and expressing the duration of the turnover of the total advanced money capital with the duration of a labour period representing a measuring unit to Marx's view, where the calendar year as a period of production in agriculture is a natural basis for this measure, the measuring unit in both cases is clearly a functional category defined by the temporal dimension of the corresponding natural, technological, and, finally, economic processes realised in the course of this unit. This demonstrates that the basis for the measurement unit for the time in which economic activity takes place is defined by internal properties of that activity or one of its component parts. Generally, this unit is neither strictly coincident nor compatible with the calendar units of measuring time. In practice, this unit is always expressed as a sum of calendar time units, but, in terms of economic activity, it uses functional aggregation to merge internal durations into the appropriate periods. Relative to their "environment," these periods represent compact and homogeneous functional units of time. In this context, whether the main measuring unit is a labour period, as presented in this analysis, or a period of production in agriculture, as used by Marx, is inconsequential.

The approach presented in this analysis differs from that used by Marx because Marx used the period of production in agriculture as a general measuring unit of the turnover time of the total advanced individual capitals, meaning that this a common measuring unit of time flow and of the temporal dimension of economic activity of all economic subjects, regardless of the actual durations of production periods or labour periods that the individually advanced money capital passed through. In addition, by making a production period of the specific type of product the general measuring unit, with its duration primarily defined by the natural properties of this type of product, Marx used this duration as a fixed length of time that remained unchanged in all circular flows of the reproduction of money capital advanced in agriculture. He does not take into account the possibility that this duration might shorten with the development of the production forces.

Drawing from these two aspects—a fixed duration and using the general character of a production period in agriculture as a measuring unit for the turnover time of all individual capitals—Marx aggregated the values and structures of all individually advanced money capitals with this time unit. Next, he analysed the causal, structural, and equilibrium relations among the individual values of economic categories, defined by this type of aggregation and formed in the course of the economic activity of the society.

In contrast, this analysis starts by assuming that because the duration of the labour periods in reproduction of individually advanced money capital is generally different and varies in different stages of development of individual capitals and of the total capital in a society, each individual capital in the certain stage of development has a functional measuring unit of its turnover time. In line with that, the duration of its reproduction time is also different.

When individual capital is advanced into the production of commodities that are of the same type in terms of use, but are different in terms of other characteristics such as labour periods, the measuring unit of the turnover time of this capital represents a weighted average of the duration of labour periods of these commodities. The turnover time and time of reproduction of the individually advanced money capital expressed in the units of a labour period that capital passes through may be called the functional reproduction time.

Different durations of labour periods as a measuring unit of the turnover time of individual money capital paid in advance and the resulting differences in functional reproduction times of such capital imply that:

- Equal calendar durations of the turnover time of two or more individually advanced money capitals is not functionally equal if measured in units of their labour periods.
- In the case of equal calendar durations of their periods of excess production period over labour period and their periods of trade or the difference in the sum of their durations not matching the difference in the duration of the labour period (with the opposite sign), the calendar duration of the periods of a single circulation flow of the capital reproduction cannot be the same.

In terms of the reproduction of an individual money capital paid in advance, this means the same calendar time unit may have different functional densities since in the course of equal calendar durations of turnover of two or more individual money capitals paid in advance, a different number of the circular flows of their reproduction may be realised.

An individual capital unit cannot be reproduced autonomously and has to establish the value and purpose relationships with other capital units within the process of exchange. For normal capital reproduction and, therefore, for establishing the equilibrium within a system, at the moment of exchange these capital units should be in the appropriate form, whether cash or commodity, as well as of the appropriate size or value. Taking into account the different functional time flows of the individual capital reproduction, a constellation of the causal and structural relationships between the economic categories in the system must be created to meet this requirement. By comprising a functional difference of the time attributes of the origin, existence, and reproduction of the certain categories, such a constellation would result in a simultaneous establishment of both content-related and time-related equilibrium of the system. In other words, the constellation establishes the equilibrium in a system in the functional time of reproduction of its component elements.

Although Marx identified the precise influence of different durations of the labour periods in a single circular flow of the capital reproduction on the modalities of this reproduction, his analysis did not extend to include the equilibrium relations among economic categories and the constellation of the causal and structural relations within and among the subsets traversed by the money capital that was advanced in the first and second sector. Starting from the assumption that the overall turnover times of these capitals have equal durations and that within this time unit only one circular flow of reproduction of the capitals advanced in the first and second sector unfolds, Marx simply neglected to include a temporal component in his analysis of the problem of creating an equilibrium in a system.

By including a temporal component, this analysis attempts to examine the relevance of this aspect on capital reproduction and on the overall economic activity in a society. The analysis proceeds from the discussion on analytical derivatives presented in Chaps. 9 and 10. For simplicity, some additional assumptions are included:

- The duration of the labour periods in the first and second sector are equal to the duration of one circular flow of reproduction in these sectors. This means that in the first and second sector, the duration of the period of excess production time over labour period and period of trade equals zero.
- The size of the total money capital advanced into the first or second sector, or both, is fixed and determined outside of the sector.
- Only simple reproduction is considered.
- The relative size of capacity expressed in the units of labour period is a constant figure defined by technical properties of the labour instruments.
- The quantity or value of the means of production that must be consumed to produce a unit of commodity in the unit of labour period is constant.
- The size of the total initially advanced money capital is constant.

In such a system labour periods, turnover times of the total advanced capitals, and the periods for which wages are paid in the first and second sector are of different durations. As a result, all temporal attributes of the circular flows of the capital reproduction in these sectors in this system are variable and different. Indeed, the length of the period needed to create a cumulative equilibrium in such a system no longer has a fixed calendar duration that is determined by the duration of initially employed labour instruments in the first and second sector. Instead, it is defined by the interaction of functional time flows in which the capital reproduction takes place in the first and second sector. (See Appendix D for an analysis of the modalities for creating an equilibrium in such a system.)

In such a system, reproduction takes place under the following conditions:

- The labour period in the second sector is v times shorter than the labour period in the first sector.
- The period for which the wages are paid in the first sector is $x(I)n(I)$ times longer than the labour period in the first sector, while the period for which the wages are paid in the second sector is $x(II)n(II)$ times longer than the labour period in the second sector. Here it is not necessary that $x(II)n(II) = vx(I)n(I)$ or that the wages in the first and second sector are paid for a calendar period of the same length.
- The length of the overall turnover of money capital advanced into the first sector is $y(I)n(I)$ times longer than a labour period in the first sector, while the length of the overall turnover of money capital advanced into the second sector is $y(II)n(II)$ times longer than a labour period in the second sector. In this case, $y(II)n(II) = vy(I)n(I)$ only when the calendar time of duration of labour instruments is equal in both sectors, which represents an atypical rather than a typical case.
- Considering that generally $y(II)n(II) \neq vy(I)n(I)$, the cumulative equilibrium of a system is established for a period defined as the least common multiple (*N.Z.V.*) of the numbers $y(II)n(II)$, $y(I)n(I)$, and $1/v$, as expressed in the units of labour period in the second sector, that is of the numbers $y(II)n(II)$ and $y(I)n(I)$ as expressed in the units of labour period in the first sector. Therefore, the length of this period is:

$$N.Z.V.\left(y(II)n(II),\, y(I)n(I),\, 1/vn(II)\right)$$

$$\text{Or}, N.Z.V.\left(y(II)n(II),\, y(I)n(I)n(I)\right)$$

In other words, this period is bounded by the moment in which the ends of the overall turnovers of money capitals advanced into the first and second sector overlap or, in other words, when the utilised labour instruments of the first and second sector must be replaced simultaneously. Thus, this period contains:

$$\frac{N.Z.V.\left(y(II)n(II),\, y(I)n(I), \frac{1}{v}\right)}{y(II)n(II)}$$

of the overall turnovers of money capital advanced into the second sector and:

$$\frac{N.Z.V.\left(y(II)n(II),\, y(I)n(I)\right)}{y(I)n(I)}$$

of the overall turnovers of money capital advanced into the first sector.

- To establish equilibrium in a system in accordance with the previous condition, the distribution of the total advanced money capital to the money capital of the first and the money capital of the second sector is defined by the expressions:

$$N(I) = \left(1 + k + \frac{na' + k}{nw(I)}\right) SR(I)$$

$$N(II) = \left(1 + k(II)\frac{1}{v}n(II) + \frac{x(II)n(II)k(II)n(II) + \frac{1}{y(II)}n(II)}{w(II)n(II)}\right.$$

$$\left.\cdot \frac{y(II)n(II)(1+se)\left(k(I)\frac{1}{v}n(II) + \frac{1}{y(I)}\frac{1}{v}n(II)\right)}{x(I)\frac{1}{v}n(II)w(I)\frac{1}{v}n(II)\left[y(II)n(II)k(II)n(II)+1\right]}\right) \cdot SR(I)$$

Considering a possibility that $y(I)n(I)$ and $y(II)n(II)$ do not represent periods of equal calendar duration and a possibility that these periods are represented by neither $x(I)n(I)$ nor $x(II)n(II)$, including a general assumption that $x(I) \neq 1$, the organic composition of capital of the first sector is defined as:

$$w(I)n(I)=x(I)n(I)\left[k(I)n(I)+\frac{1}{y(I)}n(I)\right]\frac{SR(I)}{RS(I)}$$

At the same time, the organic composition of capital of the second sector is defined as:

$$w(II)n(II)=x(II)n(II)\left[k(II)n(II)+\frac{1}{y(II)}n(II)\right]\frac{SR(II)}{RS(II)}$$

- The value of production of the first sector in a single working period of the first sector is defined as:

$$W(I)=\left[\frac{1}{y(I)}n(I)+k(I)n(I)+(1+se)\frac{k(I)n(I)+\frac{1}{y(II)}n(I)}{w(I)n(I)}\right]\frac{1}{A+BC}N,$$

At the same time, the value of production of the second sector in a single working period is defined as:

$$W(II)=\left[\frac{1}{y(II)}n(II)+(1+se)\frac{k(II)n(II)+\frac{1}{y(II)}n(II)}{w(II)n(II)}\right]\frac{C}{A+BC}N$$

(A detailed explanation of the variables A, B, C, and N is found in Appendix D.)

- The cumulative value of the total production for a period for which the cumulative equilibrium is being created in a system is defined as:

$$\sum^{N.Z.V.\left(y(II)n(II),\, y(I)n(I),\, \frac{1}{v}\right)} W = \sum^{N.Z.V.\left(y(II)n(II),\, y(I)n(I),\, \frac{1}{v}\right)} W(II) + \sum^{N.Z.V.(y(II)n(II),\, y(I)n(I))} W(I)$$

where $N.\ Z.\ V.\ (y(II)n(II),\ y(I)n(I),\ 1/v) = N.\ Z.\ V.\ (y(II)n(II),\ y(I)n(I),\ y(I)\ n(I))$.

Notes

1. Karl Marx, *Capital. Vol. 2, The Process of Circulation of Capital* (Moscow: Progress Publishers, 1956) Chapter 7 The Turnover Time and the Number of Turnovers.
2. Ibid.
3. Ibid., 72.

References

Baumol, William J. 1959. *Economic Dynamics: An Introduction*. New York: Macmillan.

Marx, Karl. 1956. *Capital. Vol. 2, The Process of Circulation of Capital*. Moscow: Progress Publishers.

Shackle, George Lennox Sharman. 1958. *Time in Economics*. Amsterdam: North-Holland.

Theil, Henri. 1974. *Linear Aggregation of Economic Relations*. Amsterdam: North-Holland.

12

The Relative Character of Functional Time

The analysis so far has examined a general course of the economic activity in time by completely excluding the calendar dimension of the time flow. All variables and parameters in the model were expressed in the units of labour periods, that is, as a function of the duration of these periods or functional measuring units of time in a system. This leads to the observations:

- Changes in duration of labour periods, of the turnover time of the total advanced capital, and of the time for which wages are paid expressed in units of labour periods also change the cumulative causal and structural relations among the economic categories in a system, as well as the modalities of creating its equilibrium.
- All this points to a direct influence of the time component on the total modalities of the economic activity of a society. This influence results from the time component being functionally integrated into the economic activity of certain subjects and society, existing within this activity and depending on the totality of characteristics of this activity. This forms a direct interdependence between time and all other elements of economic activity and creates direct mutual interaction

Ž. Rohatinski, *Time and Economics*, DOI 10.1007/978-3-319-61705-3_12

among them. As a consequence, time is an equal factor and not a mere externality for the course of the economic activity.

- Each sub-system in an economic system is reproduced in its own functional time, expressed in the units of labour periods of different duration. The relative movement between the sub-systems results in different modalities of functioning of the system and its segments. In a material-financial sense, this reflects on the temporal attributes between supply and demand, mode and form of establishing equilibrium, and the efficiency of a sub-system and the entire system in space and time.

The question of the system's efficiency in space and time opens another area of research in the relations between time and all other elements of economic activity because these elements also influence the formation of functional time in economics. At this point, a modified form of the Lorentz transformation can be applied on the functional time of the economic activity and this time can be treated in the context of the theory of relativity of time and space, with a direct reflection on the analysis of the efficiency of an economic system.

Such an approach was used by Croatian economist Branko Horvat, who analysed the problem of a contraction of costs of fixed capital and a "dilation" of economic time in steadily growing economic systems.[1] To simplify the approach, Horvat introduced the following assumptions into the model:

(a) No technological progress.
(b) Production is proportionate to the size of the fixed capital, that is, the size of capacity, so that $Y_t = pK_t$ where the production coefficient, p, is constant.
(c) Only one type of product is produced and used for all purposes.
(d) Changes in the costs of labour power and other non-capital costs are not taken into account.
(e) The value of a decommissioned machine is zero.
(f) Investment activation period is zero.
(g) Capital goods are perfectly divisible and adjustable.

Horvat defined the actual cost of fixed capital as an investment necessary to maintain the production capacity unchanged during a unit period and set the function of the capital gross fund and materialised gross investments as:

$$K_t = \int_{t-n}^{t} e^{n\tau} d\tau = \frac{1}{r} e^{n(t-n)} \left(e^{rn} - 1\right)$$

And the equation of replacement of the initially advanced fixed capital as:

$$R_t = G_{t-n} = e^{n(t-n)}$$

where r = finite constant rate of growth in the system and n = duration of fixed capital.

In this way, Horvat concluded that the actual cost of the fixed capital per unit of this capital in a steadily growing economic system can be defined as:

$$k = \frac{R_t}{K_t} = \frac{r}{e^{rn} - 1}$$

The actual cost depended solely on the growth rate of the system and the duration of the fixed capital, which makes it time-independent.

In contrast, Horvat showed that in a stationary system of simple reproduction where $r = o$, the actual cost per fixed capital unit was:

$$K^O = \lim_{r \to 0} k = \frac{1}{n}$$

From here, the following relation was formed between the unit costs of fixed capital in a steadily growing and stationary system:

$$\frac{R}{R^O} = \frac{k}{k^O} = \frac{rn}{e^{rn} - 1} = \beta$$

This relation represented a transformation factor that changes the cost of fixed capital in a stationary system into the cost of fixed capital in a steadily growing system. That phenomenon is analogous to the relativity factor in physics.

Since the assumption included a linear spending of capital within its duration, the factor β may be expressed as:

$$\beta = \frac{k}{k^O} = \frac{R}{K} \div \frac{R^O}{K^O} = \frac{1}{n} \div \frac{1}{n^O}$$

Where n^O signified the number of the calendar time units of expected duration of fixed assets in the stationary system, while n represented a number of calendar time units of duration of fixed assets in a dynamic system, that is, in a system that grows relative to the system S^o.

Therefore, $n = \frac{1}{\beta} n^o$, which, according to Horvat, meant that the effects of growth on the costs of capital are such that it seems that n_o of stationary years extends $1/\beta$ time in n dynamic years. By replacing the expression n^O with $(t_2^o - t_1^o)$ and n with $(t_2 - t_1)$, Horvat presented the dilation of economic time:

$$\Delta(t_2 - t_1) = 1/\beta\left(t_2^O - t_1^O\right)$$

From this, the time interval $(t_2^o - t_1^o)$ did not represent the calendar duration of the fixed asset, n^o, but rather the economic time of the fixed asset. Horvat wrote that the notion of "economic time" referred to a weighted average of duration of all parts of the fixed asset.

Based on this, Horvat concluded that the increase in growth rate of the system is equivalent to the creation of time, implying that every economic system has its own, specific, economic time.[2]

Comparing the assumptions of the model of expanded capital reproduction presented in this book to the assumptions used by Horvat to analyse the costs contraction and time dilation in steadily growing economic systems shows the assumptions *a, b, d, e, f,* and *g* are identical, which leads to a series of conclusions:

$$W = f\left(k, se, w, x, y; PR, RS, SR\right), \text{where the parameters are constant.}$$

$$W = pSR, \text{ where } p = k + \left(1 + se\right)\frac{k + \frac{1}{v}}{w} + \frac{1}{y}$$

$$\frac{RS}{x} = P_r, \text{ where both are constant}$$

$$\sum_{1}^{Y}\frac{1}{y}SR - SR = 0$$

$$A\left(SR\right)_{n=1} = \Delta SR_{n=2}$$

Assumption *c* on the type of product produced is not fulfilled because this analysis concerns a two-sector system. Instead, this analysis uses an approach where the entire system consists of the first sector, with production growing steadily at a finite and constant rate, *r*. (This is shown in Appendix B.)

In such conditions under simple reproduction, the value of production of a system in a single labour period, *t*, is defined as:

$$W_t = \left[k + \left(1 + se\right)\frac{k + \frac{1}{v}}{w} + \frac{1}{y}\right]SR_t$$

Or, expressed otherwise:

$W_t = p\ SR_t$, where $t = 1, 2, \ldots$ Assuming continuous accumulation in a system at a constant rate, s, so that, according to expression (13) presented in Appendix B, the size of the accumulation formed at the end of the period t of the circular flow of capital reproduction is:

$$A_t = s\frac{W_t}{w+1+se},$$

Further, assuming the investment activation period is zero at the point of transfer into the following circular flow of the reproduction, the system will employ additional production factors in values defined by the following expressions for distribution of money capital, $N(I)$:

$$\Delta SR_{(t+1)} = \frac{A_t}{A}, \text{where } A = 1 + k + \frac{x\left(k+\frac{1}{y}\right)}{w}$$

$$\Delta PR_{t+1} = k\Delta SR_{t+1}$$

$$\Delta RS_{t+1} = \frac{x\left(k+\frac{1}{y}\right)}{w}\Delta SR_{t+1}$$

Thus, production growth in the labour period $(t+1)$ in relation to production value in the labour period is:

$$\Delta W_{t+1} = W_{t+1} - W_t = \left[k + (1+se)\frac{k+\frac{1}{y}}{w} + \frac{1}{v}\right]\Delta SR_{t+1}$$

Considering that $\Delta SR_{t+1} = \frac{sW_t}{A(w+1+se)}$, the total production value in the labour period $(t+1)$ may be expressed as:

$$W_{t+1} = pSR_t\left(1+\frac{sp}{A+(w+1+se)}\right)$$

When $t \leq y$, that is when the replacement in kind of the initially advanced labour instruments ($SR_t = SR_1$) has not yet been carried out, the production value in the labour period t, where $t=1, \ldots y$, is:

$$W_t = (1+r)^{t-1}\, pSR_1$$

where r is the constant rate of production growth in the system:

$$r = \frac{sp}{A(w+1+se)}$$

At the same time, since the increase in the value of labour instruments is a linear function of the value of production in the system from the previous circuit of capital reproduction, by including two expressions:

$$\Delta SR_t = \frac{s}{A(w+1+se)} W_{t-1}$$

$$W_t = (1+r)^{t-1}\, pSR_1$$

The value of labour instruments in the system grows towards the expression:

$$SR_1 = \left[1+\frac{sp}{A(w+1+se)}\left(1+(1+r)+(1+r)^2+\ldots+(1+r)^{t-2}\right)\right]$$

It follows that the value of labour instruments employed in the labour period t is defined by the expression:

$$SR_t = (1+r)^{t-1} SR_1$$

Therefore, the value of labour instruments in the system grows at the same constant rate, *r*, as the value of production.

This also means that when there is no technological progress and the parameters *k* and *w* are constant values, the efficiency of fixed assets is equal in both simple and expanded reproduction. It follows that the unit costs of production in the system of simple reproduction are the same as in the system of expanded reproduction.

However, when the costs in a system of expanded reproduction, *S*, "moving" at the relative velocity *r* in relation to the system of simple reproduction, are viewed from the aspect of the system of simple reproduction, S^o, which is "stationary" relative to the system *S*, it leads to a "contraction" of costs growing in accordance with the coefficient β.

This is because the unit costs of production on the basis of initially advanced labour instruments, SR_1, during the turnover time of the totally advanced capital in the S^o system are defined as:

$$T^O_{(y)} = \frac{SR_1}{yW_1} = \frac{SR_1}{yp^{SR_1}} = \frac{1}{yp}$$

While costs in the *S* system are defined as:

$$T_{(y)} = \frac{SR_1}{\sum_{t=1}^{Y} W_1}$$

Considering that $W_t = (1+r)^{t-1} pSR_1$, the expression for $T_{(y)}$ may be written as:

$$T_{(y)} = \frac{1}{p\left[1 + \sum_{i=1}^{y-1} (1+r)^{t_i}\right]}$$

It follows that the coefficient of contraction of the unit production costs on the basis of the initially advanced labour instruments in the system of expanded reproduction is:

$$\beta = \frac{y}{1+\sum_{i=1}^{y-1}(1+r)^{t_i}}, \text{ where } t_i = \left(1,2,\ldots,(y-1)_i\right)$$

Since $\sum_{i=0}^{y-1}(1+r)^{t_i}$ may, through the expressions used for the sum of the first i members of the geometric series, be expressed as:

$$v = z\frac{z^{y-1}}{z-1}, \quad \text{where} \quad z = (1+r),$$

The coefficient of cost contraction or the general factor of transformation β is:

$$\beta = \frac{y}{1+v}$$

where v represents the length of the relative motion of the system of expanded reproduction relative to the system of simple reproduction.

Analogous to the "contraction" of costs and the increase of efficiency in the system of expanded reproduction compared with the system of simple reproduction, in the first system a "dilation" of time or an "extension" of the capital turnover time in relation to the capital turnover time in the system of simple reproduction takes place according to the expression:

$$\Delta(t_2 - t_1) = \frac{1}{\beta}\left(t_2^O - t_1^O\right)$$

This means that the time required by this system to achieve the same effects of the utilisation of the total advanced money capital is $1/\beta$ times shorter than the time required in the system of simple reproduction.

Consequently, the following relation is established between the cumulative values of production in the systems:

$$\sum_{i=1}^{Y} W_i = \frac{1}{\beta} \sum_{i=1}^{Y} W_i^O$$

This corresponds to reciprocal relations between the volume of the bodies moving relative to each other at the relative velocity in the physical world.

Evidently, when two economic systems or two sub-systems within the same economic system move relative to each other, with motion expressed by a steady and finite rate of production growth in one system in relation to the other, time dilation and contraction of costs of the system reproduction, as well as the turnover time of the total advanced capital, can be observed. Although from the internal aspect of the initial system, the efficiency of the capital unit in the system remains unchanged for the entire duration of the capital turnover and it is the same as the internal efficiency of the capital unit in another system, this changes the efficiency of capital and the entire system in comparison with the efficiency of the system relative to which this system is moving.

These phenomena may be explained analogously to Einstein's explanation of time dilation and length contraction in inertial reference frames moving relative to each other at the relative velocity v and by applying a modified form of the Lorentz transformations. It should be noted that in this case there is no single constant value that would be equal in all reference frames, which in physics is the speed of light, but both elements defining the value of the transformation factor β (y and r) vary in space and time. These elements are generally different in individual systems and in individual phases of development of the same system, so in analysing time dilation and costs contraction in economics it is appropriate to assume that the two elements will be constant only during a single turnover of the total capital advanced in an individual system.

This analytical derivative of time dilation and costs contraction was made on the assumption of equal lengths of the labour periods, n, in both systems:

$$n(I) = n(II)$$

It also included the assumption that the simple reproduction in one system is actually taking place, not only relatively.

Consequently—provided that the other conditions remain unchanged—the rate of the relative growth of production in one system relative to another system, r, is identical to the rate of the absolute growth of production in this system and represents the function of the accumulation rate only, while the length of the turnover time of the total capitals advanced in an individual system expressed in the units of the labour periods of this system is uniformly determined.

However, if these assumptions are not fulfilled—and generally they are not—then all the elements that define the rate of the relative growth of production in one system in relation to the other also act as a function of relation between the internal functional time factors of the capital reproduction in an individual system. Thus, this rate becomes a function of the relation between the functional times in the two economic systems.

Considering that under such conditions the turnover time of the total advanced capitals expressed in units of labour periods of different lengths in an individual system is no longer functionally unambiguous, the factor of transformation β now represents a function of the relation among four elements: $r(I)n(I)$, $r(II)n(II)$, $y(I)n(I)$, and $y(II)n(II)$. Therefore, this factor also depends on the relation between the functional time factors of the capital reproduction in the observed systems.

For illustration, assume:

- The systems employ instruments of labour of equal value and equal calendar duration, that is, equal quantity of labour power for an equal time period.
- The relative size of capacity in the unit of calendar time is equal in both systems.

- Both systems represent expanded reproduction with an equal rate of accumulation.
- The length of the labour period in the first system is longer than that in the second system because of technology or some other factor.

Under these conditions, the flows of reproduction in an individual system take place in an initial situation represented by the expressions:

$$r(i)=\frac{s(i)p(i)}{A(i)\left(w(i)+1+se(i)\right)}$$

$$p(i)=k(i)+\left(1+se(i)\right)\frac{k(i)+\frac{1}{y(i)}}{w(i)}+\frac{1}{y(i)}$$

$$A(i)=1+k(i)+\frac{x(i)\left(k(i)+\frac{1}{y(i)}\right)}{w(i)}$$

It follows that the production in terms of units of a labour period in the first system grows faster than the growth rate of production in terms of units of a labour period in the second system.

This, however, still does not mean that the production in the first system grows relatively in relation to the production in the second system since the measuring units of the functional time of the capital reproduction in the first and second system, which are used to assess these production growth rates, are different.

To determine a growth rate of production in the second system in comparable terms of time to the growth rate of production in the first system, the coefficient of the growth of production in the second system in the turnover time of the capital advanced in the second system expressed in the units of a labour period of the second system, which is:

$$kf(II)n(II) = \left(1 + r(II)n(II)\right)^{y(II)n(II)-1}$$

must be transformed into a coefficient of the growth of production of the second system in the turnover time of the capital advanced in the second system expressed in the units of a working period of the first system:

$$kf(II)n(I) = \left(1 + r(II)n(I)\right)^{y(I)n(I)-1}$$

Considering that:

$$kf(II)n(II) = kf(II)n(I)$$

This would be:

$$\left(1 + r(II)n(II)\right)^{y(II)n(II)-1} = \left(1 + r(II)n(I)\right)^{y(I)n(I)-1}$$

It follows that:

$$r(II)n(I) = \left(1 + r(II)n(II)\right)^{\frac{y(II)n(II)-1}{y(I)n(I)-1}} - 1$$

This shows the "relative motion" of one system to the other. A relative velocity of this motion is:

$$r = \frac{1 + r(II)n(I)}{1 + r(I)n(I)} - 1$$

This enables a general form of the transformation factor, β, to be expressed as:

$$\beta = \frac{yn^0}{1 + \left(\frac{1 + r'n^0}{1 + r^0 n^0} \frac{\left(\frac{1 + r'n^0}{1 + r^0 n^0} \right)^{yn^0 - 1} - 1}{\frac{1 + r'n^0}{1 + r^0 n^0} - 1} \right)}$$

where:

$$yn^0 = yn \frac{n}{n^0}$$

$$r'n^0 = \left(1 + r'n\right)^{\frac{yn-1}{y^0 n^0 - 1}} - 1$$

These final expressions show the influence exerted by the relation between the functional time flows of the capital reproduction in the specific systems on the speed of their motion and, in turn, on the factor of costs contraction and time dilation in the system moving relative to a relatively stationary system.

* * *

Summarising the set of the presented modalities of the course of the economic activity, four conclusions can be drawn:

First, each economic system as a subset of the macroeconomic system is reproduced in its internal or functional time, while the base and a measuring unit of this time is the length of a labour period within this system. This measuring unit is determined by the set of natural, technical, and economic conditions, representing the objective or relatively objective factors for the given economic system in the specific stage of development.

All of the time features of the reproduction flow in the system, such as the period of a single circular flow of the capital reproduction, the period for which the wages are paid, and the duration of the initially advanced means of labour, expressed in the units of the duration of a labour period represent the functional time flow of the system reproduction or the functional time of the system. It is imminent to the origin, existence, and termination in the certain manifestation of any economic category or variable in the system and implicitly built into all causal and structural relationships or parameters between those categories.

Second, the duration of the individual phase or process in the course of economic activity of a system is determined by the sum of the consecutive functional time units, in which the entire time-space pattern of this process is realised. This length is absolute from the aspect of the time measuring system in the specific system, but relative in terms of relationships between the time measuring systems in the systems moving at a uniform speed.

This means that in a system moving relative to a stationary system, a time dilation can be observed. Dilation may be analytically described by applying a modified form of the Lorentz transformations. In that case, the transformation factor represents the function of speed of the system, turnover time of the total capitals advanced in the systems, and the duration of the labour periods in an individual system. This is also related to the issue of different efficiency of the systems in the same "non-temporal" conditions of reproduction.

Third, normal reproduction of two or more functionally related economic systems in different functional times requires an integral material-financial-time equilibrium of the macroeconomic system to be established. The relationship between the functional reproduction time flows of individual systems directly affects the allocation and structure of the use of capital in the macroeconomic system, the "equilibrium" size of the certain economic categories or variables in each system, structural and causal relationships or parameters among these categories in an individual system, absolute and relative efficiency of each individual system, and the time modalities of establishing a macroeconomic equilibrium. As a result, functional times of individual systems are the active factors that initially and recurrently affect the total forms of reproduction in these systems and in the entire macroeconomic system.

And finally, the functional unity of functional times of every system within the macroeconomic system represents its functional time. Since the macroeconomic system comprises numerous systems that differ in size and complexity, in their hierarchical levels, and in activities that follow, overlap, and oppose one another, the functional time of the macroeconomic system is difficult to view empirically. Therefore, in practice, the time flow of its economic activity is usually expressed, analysed, and measured in calendar time units.

However, from the aspect of economic activity of an individual system and the macroeconomic system as a whole, calendar time units are generally non-functional. Therefore, a compulsory temporal aggregation of such activities into uniform calendar time units results in losing insights into their actual or functional flow, while also creating a partially distorted image of the efficiency of such activities.

This distortion stems from three related reasons:

- Because of different durations for labour periods, periods of circular flow of capital reproduction, and the turnover time of the total advanced capital in the individual systems, the same calendar time unit contains different amounts of the functional time units of an individual system and so a different number of economic events and processes representing a functional part of one or more circular flows of reproduction of the total capital advanced in an individual system. Additionally, these calendar time units have different densities, as well as variances in their functional scope and in their content. Unambiguous aggregation of such different contents into a fixed calendar time unit obscures a comprehensive view of the functional flow of the system reproduction within the calendar time unit.
- Such an aggregation sums up the values of the variables within the individual systems and in the macroeconomic system, which functionally pertain to different circular flows of capital reproduction and turnovers of the total advanced capital. The actual value of parameters among the variables within the system, which is functionally established in the unit of the circular flow of reproduction, is distorted.
- Calendar time units transect this flow in an individual system in different ways and in different functional phases of the flow of reproduction.

This distorts the perception of actual functional relationships between the systems and also the perception of the modality of functioning of the macroeconomic system.

From this, expressing the temporal dimension of economic activity in the calendar time units may only to a limited extent simulate the actual time flow of the economic activity of an individual subject or system and a society. However, this time flow is actually realised in the functional time of each individual system and in the functional time of the macroeconomic system as their functional unity.

Notes

1. Branko Horvat, "Certain Similarities between Inertial Systems in Physics and Steadily Growing Systems in Economics," *Ekonomska analiza* 197, no. 1–2 (1987): 47–58.
2. Branko Horvat, *Radna teorija cijena* (Belgrade: Izdavacka radna organizacija "Rad," 1987) 163–165.

References

Horvat, Branko. 1987a. Certain Similarities between Inertial Systems in Physics and Steadily Growing Systems in Economics. *Ekonomska analiza* 197 (1–2): 47–58.

———. 1987b. *Radna teorija cijena i neki drugi nerijeseni problemi ekonomske teorije*. Belgrade: Izdavacka radna organizacija "Rad".

13

The Time Determination of Economic Activity

With this understanding of the time component of the economic activity, the analysis progresses to the "time determination" of this activity, in other words classifying the time periods of various lengths depending on the characteristics and modality of the economic activity in each one of them. Marshall's ideas on the modalities of creating the equilibrium between supply and demand on the commodity market are directly relevant to this question.

The English economist believed these modalities represent the functions of different levels of flexibility of supply and demand in the unit of time. In his 1890 treatise, *Principles of Economics*, where he attempted to harmonise the labour theory of value of the classical political economics with the concept of marginal utility, Marshall was the first to focus on this relatively neglected area within classical political economics. In his book, he developed the analytical instruments for the analysis of how time influences economic activity, and many of the elements he included have outlived the other theoretical assumptions of their author.

The unfolding of the economic activity of certain subjects and society in time periods of different duration has been analysed in classical political economics primarily with the objective of defining key factors determining the size, structure, and relations between supply and demand of

Ž. Rohatinski, *Time and Economics*, DOI 10.1007/978-3-319-61705-3_13

the production factors and commodities on the market, and, from there, the mode of setting prices in the specific time period. Within this framework, classical political economics strictly distinguished the normal or natural prices or values of products formed over the long term and defined by the values of the relevant economic categories in the sphere or phase of production from the short-term market prices defined by the relations between the relevant economic categories in the phase of exchange.

The important elements of this approach were included in Quesnay's earlier analysis of the differences between the *prix naturel* acquired by a product in the process of production and manifested when "selling the product first hand" or at the moment the product is transferred from the sphere of production to the sphere of exchange and *prix courant*, which is determined by the relation between supply and demand for this product on the market.[1] But it wasn't until 1776, in Adam Smith's *Wealth of Nations*, that these differences were presented comprehensively as a basis within the labour theory of value.

Smith first defined the price of commodities as a value determined by natural or long-term prices of the employed production factors of land, capital and labour. Next, he analysed the formation and size of short-term market prices in conditions of perfect competition, as well as their relation towards the natural price.

The pioneer in political economics concluded, "The market price of every particular commodity is regulated by the proportion between the quantity which is actually brought to the market and the demand of those who are willing to pay the natural price of the commodity, or the whole value of the rent, labour, and profit, which must be paid in order to bring it thither."[2] The market price is equal to the natural price only "when the quantity brought to the market is just sufficient to supply the effectual demand."[3] The analysis of how market and natural prices are harmonised by establishing the equilibrium of supply and demand in a unit of time is especially important when considering the definition in classical political economics of the basic features of behaviour of economic categories in time periods of different duration. Smith approached this issue by partially analysing the factors of forming supply, regardless of the factors affecting the size of demand, where the first ones are variable, and the second ones are fixed.

In particular, Smith noted there was a potential for short-term changes in demand within the framework presented by the difference between potential and effective demand, which was the actual demand at a given level of the natural price, and that potential was proportionate to the price trends. Yet, he believed the driving force in forming the market equilibrium was the variations in production or supply. "The quantity of every commodity naturally suits itself to the effectual demand," he wrote.[4]

The core of Smith's view of the difference between the modalities of the short- and long-term course of the economic activity can be reduced to: The course of the economic activity in the short term is directly affected by the high level of elasticity of supply of all production factors in relation to their price trends, while the long-term economic activity is primarily defined by the actual material and human potential and objective needs for the specific products in the system. However, Smith did not analyse the duration of the time period needed for an impulse, once given, to change supply of the certain production factor or, even further, to change of production and supply of the product. Also, Smith did not consider that the elasticity of supply of each of the three production factors is different in the same unit of time, which would mean their supply is defined by the specific level of their mobility.

By providing explanations for the short-term relations of supply and demand on the market based on activity of the factors that are expected to change over the long term, Smith failed to define the specific features of the short- and long-term behaviour of the economic subjects or, in other words, the formation of values and establishing of the appropriate relations between the economic categories in the system.

Smith's perception of the temporal dimension of the course of economic activity is characteristic for all of classical political economics. In one example, while criticising the view held by James Maitland, eighth earl of Lauderdale, and Jean-Baptiste Say that the value of every commodity rises as demand rises or as supply falls, British political economist David Ricardo accepted Smith's interpretation of the activity of a mechanism of balancing the market and natural prices and also his understanding of the time determination of economic activity.

"In the ordinary course of events", Ricardo wrote in 1817, "there is no commodity, which continuous for any length of time to be supplied precisely in that degree of abundance, which the wants and wishes of mankind require, and therefore there is none which is subject to accidental and temporary variations of price. It is only in consequence of these variations, that capital is apportioned precisely, in the requisite abundance and no more, to the production of the different commodities which happen to be in demand. With the rise or fall of price, profits are elevated above, or depressed below their general level, and capital is either encouraged to enter into, or is warned to depart from the particular employment in which the variation has taken place."[5]

Such an interpretation, however, leaves open the question of adapting the quantity supplied to that demanded in consideration of the differentiated mobility of the factors of production as a function of the length of the period from the moment an impulse of demand occurs until the moment it influences the formation of supply. Neither Ricardo nor his followers succeeded in finding a satisfactory solution, which would allow the observation of the partial behaviour and generalisation of the common behaviour of producers and consumers in a period of the specific duration and would also be compatible with the basic postulates of the labour theory of value.

Marshall's Approach

One of the most prominent economists of the late nineteenth century, Marshall also tackled the problem of time determination of economic activity. He approached the problem based on views that represented a combination or compromise between Ricardo's conception of the labour theory of value and the theory of the marginal utility, which essentially represents the temporal and content-based restrictions of the law of the value as a determinant of the price formation. The starting premise of Marshall's theory was that for each unit of time on the market there was only one product price. Regardless of the duration of a period, this price was always formed in the process of balancing the supply and demand, so both factors influence price autonomously in the short term, as well as in the long term.

A difference between short- and long-term prices was not a result of the partial activity of only one factor, but stemmed from a specific combination of relations between supply and demand, with a different level of elasticity in a given unit of time. Consequently, depending on the duration of time period, either supply or demand acquired a constant or variable character. In each time period there was a specific "normal" price for that period, defined at the point of equilibrium of the price of supply and the price of demand formed in the respective period. At that point, production had no tendency to rise or fall, and "if any accident should move the scale of production from its equilibrium position, there will be instantly brought into play forces tending to push it back into that position."[6] Marshall did not fully deny the classical postulate on the absolute influence made only by the costs of production on the normal or natural price, but he reduced it to the stationary state in which "in the general conditions of production and consumption, of distribution and exchange remain motionless…and the cost of production governs value.… Each element of cost would be governed by 'natural' laws subject to some control from fixed custom…, [and] there would be no distinction between long-period and short-period normal value."[7] At the same time, the stationary state was a situation in which the natural price is generally equal to the average market price.

With these ideas, Marshall introduced into the analysis the concept of a "two-dimensional" character of the equilibrium market, reflecting that the equilibrium is not only a function of horizontally connected or static elements of the values of supply and demand, but also of their differentiated motion in time. "For indeed the supply and demand schedules do not in practice remain unchanged for a long time together, but are constantly being changed; and every change in them alters the equilibrium amount of the equilibrium price, and thus gives new positions to the centres about which the amount and the price tend to oscillate," he wrote.[8]

In elaborating the temporal aspect of the formation of supply and demand, Marshall echoed the views of the classical political economists and believed that the size of demand in every single period was defined by the elements independent from the size and structure of supply so that, in the process of establishing the market equilibrium, it was an exogenous variable. Further, there was a temporally unique law of demand showing the price of demand was inversely proportional to the size of supply, and the only difference between short- and long-term demand was a long-term increase of its elasticity in relation to supply. As a consequence, the demand trend did not provide a unique characteristic to the economic activity of the certain subjects in different time units. And further, Marshall wrote, "the unit of time may be chosen according to the circumstances of each particular problem: it may be a day, a month, a year, or even a generation; but in every case … it is to be assumed that the general circumstances of the market remain unchanged throughout this period; that there is, for instance, no change in fashion or taste, no new substitute which might affect the demand, no new invention to disturb the supply."[9]

Contrary to the autonomous tendencies of demand, supply tended to adjust to the given size and structure of the demand, but its course was limited by the total and partial rigidity of the production factors. The intensity of this rigidity was inversely proportional to the length of the observed period. Marshall felt the technological and organisational ability of supply to adjust to the demand represented the main criteria of categorisation of time periods and a determinant of the specific features of the economic activity in one of these periods. Taking this into account, he distinguished four periods: very short periods, short periods, long periods, and secular periods.

In a very short period, the elasticity of supply of the production factors generally equalled zero, so changes in the demand could not lead to changes in production and supply. The constancy of supply at the same time meant that any current variations in demand would be followed by the proportional movement of prices. The level of these prices was independent of the natural price of the production factors and the costs in the very short period. Viewed in isolation, the ultimate conclusions of this concept were not essentially different from the classical view of forming the market price. As such, they presented a precise elaboration of the process of the current balancing of supply and demand rather than an original contribution to the theory of economics. However, this concept contained and systematised all the relevant elements of Marshall's approach to the problem of dynamics in economics. Interestingly, however, when viewed in the context of contemporary dynamics, his theory was static, which meant that unlike the classical perception, it did not present an explanation for any deviations of the specific tendencies from the general rules, but instead illustrated the temporally specific activity of the unique system of forming the normal price.

Moving onward, however, Marshall's analysis of the short period represented an original approach to filling the gap witnessed in classical political economics between the market and natural price and to using this as a basis for creating a dynamic theory of economic activity. In his opinion, the main characteristic of this activity in the short term was that "the supply of specialized skill and ability, of suitable machinery and other material capital, and of the appropriate industrial organisation has not time to be fully adopted to demand; but the producers have to adjust their supply to the demand as best they can with the appliances already at their disposal."[10] In other words, the human and material economic resources were mostly fixed in the short term, while the attempt of supply to adjust to a given level of demand focused on increasing or decreasing the level of utilisation of the existing resources, through greater or lesser employment of the variable production factors. This meant that the price of supply at the point of the short-term market equilibrium was determined by the prices of the employed variable production factors, or, more precisely, it was a function of the marginal costs of production.

In contrast to the short-term inelasticity of the production capacities or of the capital and the so-called fixed production factors, Marshall believed in the long period the supply of all factors of production, including production itself, had time to adapt to the changes in demand. As a result, the normal price of the long period was determined at the intersection of the function of demand and the function of the total production costs of the certain product.

Marshall rounded up the time aspect of his theory of economic equilibrium with a fourth temporal category, the secular or very long period. Characteristics of the secular period included changes in professional knowledge, population, size, and structure of capital, as well as in the general conditions of supply and demand. This temporal category exceeded the framework of the initial premise that referred to the homogeneity of supply and demand in the observed time unit. In this context, the secular period represented a "temporal universe" rather than the object of an exact analysis.

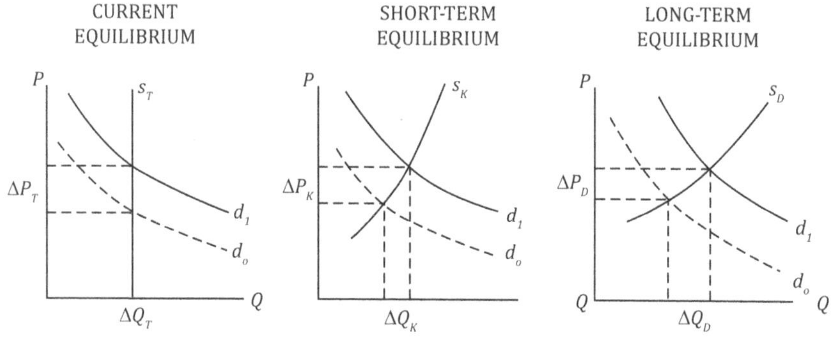

Fig. 13.1 Marshall's classification of equilibrium. Note: Q = quantity, P = price, d = demand, s = supply; $\Delta P_T > \Delta P_K > P_D$; and $O = \Delta Q_T < Q_K < \Delta Q_D$

The characteristics of movement and the mutual balancing of supply and demand in very short, short, and long periods are commonly represented across three graphs (Fig. 13.1).

Looking at the duration of these three periods, Marshall generally represented a functional viewpoint, suggesting the duration of a period was a function of the level of elasticity of a concrete supply in relation to the trends of demand, regardless of the number of the calendar units needed for harmonisation. This meant short and long periods of time may be of different calendar duration in relation to the production of different groups of products. Suggesting that the short term generally represented a period of up to one year, long term several years, and very long term about a generation was actually only a rough estimate of an average. Conscious of this aspect, Marshall concluded, "Of course, there is no hard and sharp line of division between 'long' and 'short' period. Nature has drawn no such lines in the economic conditions of actual life; and in dealing with practical problems they are not wanted."[11]

As any attempt to reduce an objective reality to fit the framework of specific conceptual categories, Marshall's classification of time periods contained a high level of arbitrariness. The problem was exacerbated by the difficulties in defining the criteria for a precise division of the production factors into fixed and variable. His interpretation that the difference between these factors derived from a different level of elasticity of their supply in the unit of time and, further, his approach to classifying time periods based on whether product supply was determined only by variable or total production factors led Marshall into a cause-and-effect cycle that was hard to escape despite his efforts to clarify the question using a descriptive method.

The essence of this problem was clearly pointed out by Charles E. Ferguson. By defining fixed factors as inputs "whose quantity cannot readily be changed when market conditions indicate that an immediate change of output is desirable," the twentieth-century US economist concluded:

> To be sure, no input is ever absolutely fixed, no matter how short the period of time under consideration. But frequently, for the sake of analytical simplicity, we hold some inputs fixed, reasoning perhaps that while these inputs are in fact variable the cost of immediate variation is so great as to take them out of the range of relevance for the particular decision at hand. Buildings, major pieces of machinery, and managerial personnel are examples of inputs that cannot be rapidly augmented or diminished. A variable input, on the other hand, is one whose quantity may be changed almost instantaneously in response to desired changes in output. Many types of labour services and the inputs of raw and processed materials fall in this category.[12]

Despite the shortcomings of Marshall's understanding of the relations between certain economic categories in the process of reproduction and its eventual abandonment, his classification of time periods and description of the main characteristics of the process of balancing supply and demand in a unit of time represented one of the foundations of the micro-economic analysis and continued to be valuable in a modified form. As US economist Joseph Schumpeter (1954) noted, it was only natural that Marshall's work became outdated, but it also represented a living source that will never become obsolete.

Yet, in view of contemporary changes in the conditions and modalities influencing the economic activity of individual subjects and society, Marshall's classification of time periods has faced increasing practical and theoretical problems. Loosely, these challenges can be described as market-related, technical-technological, and theoretical and systemic in nature, although there is a significant degree of interdependence.

The market-related factors encompassed issues around the flexibility of prices and the speed that they adjust to changes of supply and demand on the market. While Marshall presupposed "unlimited" price flexibility in the very short period, later research indicated a certain level of price rigidity in all periods. Among other factors, this rigidity was caused by an imperfect awareness of the subjects on the market, the cost of increased awareness, an increased possibility of redirecting demand towards alternative products that reflects increased competition, and—especially important—the increased ability of the economic subjects to organise themselves at the moment of entering the market of the production factors and commodities, both in supply and demand.

Better organisational ability erodes the market's presupposed "atomistic" character and relations between supply and demand in very short and short periods are no longer the exclusive determinants of price. In 1977, French economist Edmond Malinvaud wrote in *The Theory of Unemployment Reconsidered* that as societies reach higher degrees of organisation the classical belief that prices react quickly to surpluses in supply or demand was no longer especially useful in short-term macroeconomic analysis. This detachment is evinced in the analysis of individual decisions and a growing volume of knowledge about how prices are determined, he argued, explaining that while quick price adjustments might be seen in many agricultural products and raw materials, a similar move is not witnessed by the prices for manufactured goods and services or for wages.

Malinvaud explained the situation changes in the longer run when an increased demand for goods, services, or specific types of labour that cannot be mirrored by increased supply at constant real costs tended to trigger an upward shift in prices.[13] He argued this shift was primarily a result of differences in the relative magnitude of price increases when they occur, albeit infrequently. The immediate impact of any shift in demand or supply is seen in "order books, waiting lines, inventories, delivery dates, output, hours of work, employment," and similar places, he wrote, concluding that these are the first signals of changes in the supply–demand balance, while price adjustments occur later and are less obvious.[14]

The technical-technological challenges to some elements of Marshall's classification of time periods reflect both the determination of the calendar duration of a certain period and the possibility of distinguishing between the variable and fixed production factors. Because of today's rapid technical-technological development, along with the quickly shrinking time needed between the acquisition of new scientific knowledge and its deployment in practical applications, conditions for modern production process to begin and continue are more complex than witnessed by Marshall in the late nineteenth and early twentieth centuries. In addition, the absorption capacity of the market or potential demand has increased, which led to a corresponding increase in optimal supply capacity, as well as an extended assortment of supplies. Combined, these features significantly elongate the period needed to commission new capacity. Under Marshall's definition, this would extend the calendar duration of the short term when capacity was fixed from the period of a few months to the period of several years. In turn, it would lead to an increase in the amount of capital that needed to be advanced in the production process as a lump sum and remained relatively fixed in the process.

On the other hand, a higher degree of technology requires higher specialisation of all production factors. Individual factors are at times intended exclusively for the production of the certain type of products or even a single product, limiting the possibility for substitution. As a consequence, the existing market supply cannot always ensure a prompt increase in employment of all variable production factors, whether creating new assets or increasing production at existing assets. In practice, a portion of the "variable" factors is actually relatively fixed. This phenomenon extends the time required to expand production by increasing the use of variable factors with an unchanged quantity of fixed production factors. It changes the lower limit of the short term and extends the very short period.

The third group of challenges, those related to theoretical and systemic concepts, was sparked by changes in the organisation and modalities of how economic activity unfolds compared with the conditions present during Marshall's time. The evolution of these concepts has significantly affected the theoretical knowledge on the mode of functioning of the economic system and the organisational structure and activity mechanisms of the main subjects of the economic activity.

The initial basis of Marshall's theory in a narrow sense and the orthodox theories of John Stewart Mill, Arthur Cecil Pigou, and Leon Walras in a broader sense were:

- Micro-economic approach to the analysis of the reproduction and functioning of an economic system.
- Maximisation of individual benefits or profit as a presupposed motive and reason for the overall economic activity.
- Analysis of the modalities of establishing the partial equilibria on the individual markets of commodities and production factors.
- Acceptance of the Say's law.

The global Great Depression of the 1930s laid bare some of the weaknesses of these assumptions, motivating numerous economic researchers to explore the question. They were particularly interested in the inability of the market mechanism to autonomously harmonise the quantity of aggregated supply and demand under conditions of full employment of the factors of production.

Keynesian Approach

John Maynard Keynes led these new thinkers. In his opus, *General Theory of Employment, Interest and Money*, Keynes rejected Say's law and the *laisser–faire* concept that a general equilibrium was automatically created in a system in the conditions of full employment. He highlighted the possibility of partial equilibrium on the commodities and capital markets with a simultaneous excess of supply to demand on the labour market. To build such a theory and create an adequate analytical model, Keynes not only opposed the main postulates of Marshall's theory and the entire classical economic thought, but also redefined the modalities of the unfolding of the economic activity in periods of different duration.

Like Marshall, Keynes in his analysis of the time component of economic activity used a method of comparative statics. The scheme is commonly represented as a series of moments (Table 13.1).[15]

Table 13.1 Action-reaction scheme

T0	t1	t2	t3	T1
	The moment of the initial change of parameter	The moment of observing the change	The moment of taking action	
	Time lag in observing the change	Time lag in taking action		
	Time between the moment of change and the moment of taking action		Time between the moment of taking action and the moment of establishing the new equilibrium in a system	
	Time between the moment of change of parameter and the moment of establishing the new equilibrium in a system			

However, the scheme did not provide insights into the actual time path of the system from the moment the static equilibrium was disturbed to the moment the new equilibrium was created. Keynes focused on the character and modalities of the macroeconomic process the system used to adjust to a disorder that caused a general disequilibrium, and he was not especially interested in an analysis of the time path of the system. As a result, he did not examine the question of the time dimension of establishing and maintaining the equilibrium of the system on the market of commodities and capital in conditions of underemployment and employment. Again similar to Marshall, Keynes approached the analysis of the dynamic processes by analysing the modalities of the unfolding of the economic activity in the periods of irrelevant calendar duration, which were internally homogenous from the aspect of the behaviour of individual economic categories and of the system. Within this framework, Keynes paid special attention to the modalities of the unfolding of the economic activity in the short period, but he used a definition that essentially differed from the one used by Marshall.

Under Marshall's analysis, the equilibrium between supply and demand in the very short and short period was primarily dependent on changes in prices, which led to a hypothetical conclusion that the speed of price adjustments to the difference between supply and demand was "infinitely" high, while the speed of adjustment of the capital was zero. In contrast, Keynes' analysis suggested the opposite. In the short period, the sizes of supply and demand on the market varied, while the prices were relatively stable. Because of this, Keynes assumed that in the short period wages, population, capital, production capacity, and production technology were constant categories, while consumption, savings, investment, and employment were variable.

Keynes' assumption was based on the assessment that the size of income of the economic subjects was not defined by the quantity of services these subjects were ready to offer at the certain equilibrium price, but rather by the quantity that they had effectively sold and as a result their demand on other markets would be limited by the effectively achieved income. This restriction reduced the effective demand on the commodity market with a multiplicative influence on the total income and also on the level of the economic activity in a system. Keynes took the functional perspective on the calendar lengths of the duration of the short period as seen this way. In his approach the actual calendar duration of this period represented a function of the time flow through which the system must pass to enable the multiplier of consumption to establish the new equilibrium between supply and demand.

Even though Keynes was not especially interested in the long period, brushing it aside as a largely academic concept, he tried to define it to analyse the characteristics and modalities of the course of the economic activity in the short run. In his view, in a long period all parameters and values of the economic categories in the system were variable.[16]

Based on these views on the modalities of the behaviour of individual economic categories and the economic system in the periods of different duration, Keynes' model emphasised the main determinants of the formation and allocation of income and the level of employment in the short period. In the context of the time component of economic activity, only a portion of Keynes' original model is directly relevant and examined in this analysis.

Keynes' model implicitly contained a premise that in the short term the autonomous flow of investment did not lead to quantitative and qualitative changes in the physical volume and structure of capital, and because of this the analysis of the course of economic activity in the short period focused on factors which define the higher or lower level of utilisation of the existing capacities. The model elaborated on the process of creating the supply and demand, as well as their harmonisation on the markets of commodities, labour, and capital, and included a series of pertinent variables: nominal and real income, Y and y; employment, n; supply of money, M; interest rate, r; nominal and real wages, W and w, and the level of prices, P.

The real values of the individual variables represented a ratio between the appropriate nominal value and the level of prices expressed in terms of wages. Since the supply of money was taken as an exogenous variable determined by the state, the model consisted of seven unknowns. The solution of this model was defined by five functions—savings, investments, liquidity, production, and labour supply—and two identities—$Y = Py$ and $W = Pw$. In addition, the model also contained the balance sheet equation for dividing income into personal consumption and investments.

The entire model is commonly presented as the relations (numbered for later reference):

$$Y = C + I \tag{13.1}$$

$$S = I \tag{13.2}$$

$$S = S(Y) \tag{13.2.1}$$

$$I = I(r) \tag{13.2.2}$$

$$M = L_1(Y) + L_2(r,P) \tag{13.3}$$

$$Y = Y(n) \tag{13.4}$$

$$w = Y'(n) \tag{13.5}$$

$$W = W_o + W(n) \tag{13.6}$$

$$Y = Py \tag{13.7}$$

$$W = Pw \tag{13.8}$$

where C = personal consumption, S = savings, I = investment, L_1 = transaction demand for money, L_2 = speculative demand for money, and W_o = lower limit of the nominal wage.

Keynes' relations showed that income, employment, and the total economic activity were determined by the functions of consumption, investment, and liquidity. At the same time, the form of these functions was based on three "psychological laws" that represented the framework of his macroeconomic theory:

- *Consumption*: In Keynes' words, "Men are disposed, as a rule and on the average, to increase their consumption as their income increases, but not by as much as the increase in the income."[17] This meant that the marginal propensity to consume is greater than zero and less than one. On this basis, Keynes, working with Kahn, formulated a static income multiplier that showed an increase in consumption had multiplier effects on the increase of income, and the multiplier equalled a reciprocal value of the marginal propensity to save.
- *Incentive to Invest*: The level of investment activity was defined by the "relationship between the marginal efficiency of capital and the rate of interest," where the marginal efficiency of capital represented "the relation between the prospective yield of the capital asset and its supply price or replacement cost, i.e. the relation between the prospective yield of one or more units of that type of capital and the cost of producing that unit."[18] By definition, the inducement to invest lasted until the moment on the investment demand curve that the marginal efficiency of capital equalled the rate of interest.
- *Liquidity preference*: The need for cash depended on the transactions motive and the precautionary motive, on one hand, and on the speculative motive, on the other.

Along with a cursory presentation of the variables, relations, and assumptions included in Keynes' model, two main elements in the model are especially important to an analysis of the time component of economic activity, beyond his similarity with Marshall on the concept of time periods, which is evident in Eq. (13.4).

The first is that Keynes' model represented the short-term equilibrium of a closed economy. In this sense, the concept of equilibrium signified the inherent simultaneous equality of supply and demand on the markets of commodities, capital, and labour where the value of parameters and endogenous variables was determined in the manner that satisfied the main condition of the system equilibrium, which was the equivalence of ex ante investments and savings. The level of investment depended on the marginal efficiency of investment and was a decreasing function of the interest rate, while the level of savings depended on the marginal propensity to save and was an increasing function of income. The size of income at the equilibrium point was determined by the multiplier operation where, assuming the short-term autonomy of income, the marginal propensity to save oscillated depending on the level of income.

This conclusion implied that the first unknown in Eq. (13.2.1)—the marginal propensity to save—was determined from the equilibrium level of income and the corresponding level of savings, which was the difference between income and the level of consumption determined by that income when the short-term marginal propensity to consume was constant. When specifying the interest rate that would match the equilibrium level of income, it was necessary to add to the condition of equilibrium Eq. (13.3), which determined the money market equilibrium when the given supply of money, M, matched the transactions and the speculative demand for money, the transaction demand was an increasing function of income, and the speculative demand was a decreasing function of the interest rate, but an increasing function of price.

This determined the relations between the level of income and the interest rate. Income was a decreasing function of the interest rate, which meant that as the rate of interest tended towards infinity, the speculative demand for money tended towards zero and the income asymptotically approached the maximum value Y which could be funded by the given supply of money, M. At a given price level income was an increasing function of the interest rate, which meant that as the interest rate decreased towards its lower natural limit, r^{-}, the speculative demand tended towards infinite elasticity by employing the entire supply of money, while the transaction demand and income tended towards zero. This system of two equations and two unknowns had a unique solution for the level of income and the rate of interest at the intersection of the curves of the equilibrium movement on the commodity market, the IS curve, and money market, the LM curve, depending on the appropriate combinations of the level of income and the rate of interest. As the coordinates of their intersection were set by the given level of prices, the equilibrium levels of income and interest rate could be expressed as a function of the level of price.

The demand for labour power on the labour market is defined by Eqs. (13.4) and (13.5). The first is a function of production, which showed the level of production or income was an increasing function of the number of employees, and the second shows real wages depended on the decreasing marginal productivity of a given quantity of employed labour, n. Integrating these two equations created a unique function of demand for labour power in which demand represented a decreasing function of the real wages. On the other hand, labour supply was an increasing function of nominal wages, which became inflexible as it approached a specific minimum level, W_o. The minimum level of wages corresponded with the minimum level of employees, n^{-}, who would work that wage rate. With a given level of prices, P, connecting the nominal and real wages, the equilibrium created at the intersection of the functions of supply and demand determined the equilibrium level of employment and real wages. Like the equilibrium of the level of income and interest rate on the commodity market and money market, the coordinate of the intersection of labour-power supply and demand, for example the appropriate level of employment and wages, was a function of the given level of prices.

Based on this definition of the partial equilibria on the individual markets, the general equilibrium in Keynes' model was created by determining the level of employment that produced an equal level of the value of income at the point of equilibrium on the commodity market and money market, on one hand, and on the labour market, on the other. (The labour market equilibrium could be the result of the production function.)

In this context, all variables depending on the employment level must be expressed in units of employment. This step reduced the system of Eqs. (13.2)–(13.6) to three, which determined the equilibrium on the commodity market (13.9), money market (13.10), and the supply of labour power (13.11) in a new way:

$$S\left(y\left(n\right)\right)=I\left(r\right) \tag{13.9}$$

$$M = kPy(n) + PL(r) \tag{13.10}$$

$$Py(n) = W_o + W_{(n)} \tag{13.11}$$

Substituting the units of income and interest rate for the units of employment led to the final solution for this model.

The general equilibrium of such a system may be presented by a diagram with its coordinate axes representing the equilibrium sizes of income, employment, wages, and interest rate (Exhibit 8). The first quadrant shows equilibrium on the market of commodities and money, the second represents the production function with a decreasing marginal productivity of labour, and the third shows the equilibrium on the labour market (see Fig. 13.2).

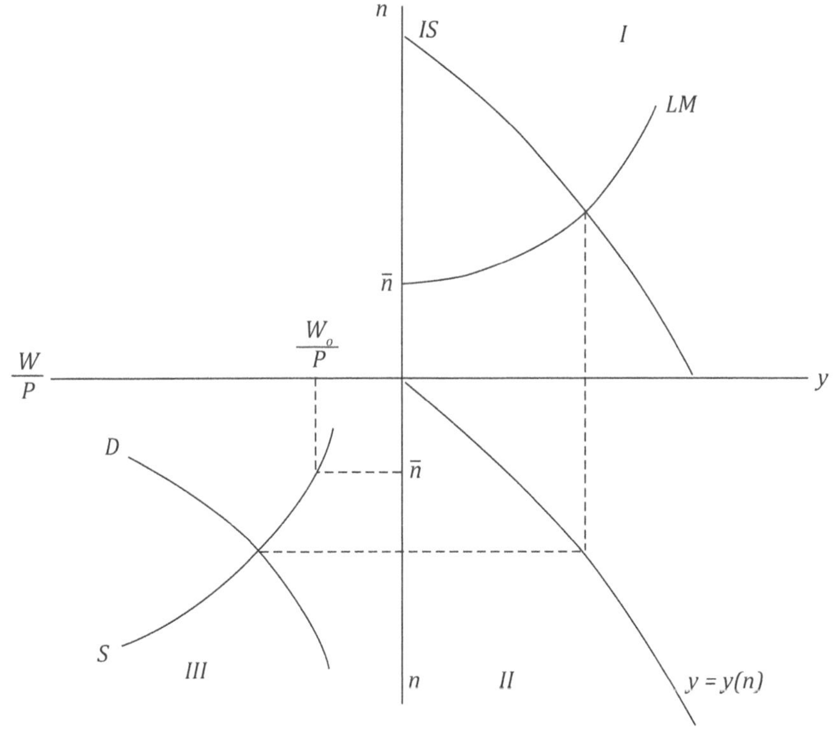

Fig. 13.2 The Keynesian model

The second key element of the model in this analysis was the three essential differences between Keynes' views on the determinants of the formation of supply and demand on the markets of commodities, capital, and labour and those of the classical school, which were essentially:

- Savings primarily represent a function of income, not of the rate of interest.
- Demand for cash, apart from being dependent on the level of income, depends on the speculative demand, for instance related to the level of interest rate.
- Labour supply depends on nominal wages, not real wages, and wages have a lower limit, below which supply becomes inelastic.

The assumption on labour supplyis perhaps most import to this analysis. It represented a direct negation of the universality of the classical postulate that held "the utility of the wage when a given volume of labour is employed is equal to the marginal disutility of that amount of employment."[19] Keynes reduced this concept to merely a specific case of full employment and set the fundamental premises for the main conclusion of his General Theory: that involuntary unemployment may exist even when the system is in equilibrium.

By defining the function of labour supply, Keynes continued to analyse the trend of the function of demand and the position of the point of equilibrium on the labour market as dependent on the trend of the total supply and demand of commodities, as related to the size and distribution of income, expressed in the units of employment, with the production function (4) representing the total supply, and the balance sheet Eq. (13.1). Total demand was represented as:

$$\phi(n) - \lambda(n) = I \tag{13.12}$$

From here, employment depended on the function of the total supply, $y(n)$, where in the state of equilibrium, $y(n) = \phi(n)$, with the propensity to consume, Λ, and with the value of investment, I, each individual level of employment has a unique corresponding value of production or income, while the marginal value of this production determined the real value of wages. The equilibrium level of employment was determined at the intersection of the functions of the total supply and demand, at the point where:

$$Y(n) = \lambda(n) + I \quad (13.13)$$

If the total supply increased in the short run following the higher employment of labour power with fixed capacity, the value of the real income, /*y* (*n*)/, also increased, and so in line with the marginal propensity to consume, the consumption of the population, /*Ƙ* (*n*)/, would increase as well. The remaining portion of income, *y*(*n*) – *Ƙ*(*n*), would appear as savings. The crucial question related to establishing the equilibrium on the commodity market in the circumstances of full employment was whether the given volume of savings would result in the required increase in savings, given the multiplier, that was needed to equalise the total demand with the resulting level of the total supply.

The solution depended on the relation between the marginal efficiency of capital and the interest rate or, in other words, on the inducement to invest and the liquidity preference. If, under the influence of these elements, the level of investment was not sufficient to raise the effective demand to match the level of supply at full employment, the result would be a corresponding decrease in the total supply and, in turn, in employment, below the level of supply of labourers who want to work at the current real wage. The marginal disutility of the equilibrium level of employment would fall under the real wage, and the system would achieve the new equilibrium with the existing unemployment.

In the diagram of the Keynesian model, the insufficient effective demand, ceteris paribus, would manifest itself in a shift of the IS curve to the left and therefore reduce the equilibrium level of income. This would in turn trigger a shift of the function of demand on the labour market to the right, causing it to intersect the function of supply in its inelastic zone by determining the point of equilibrium with the coordinates (W_0, n) where $o < n' < n$. The difference between the number of labourers, *n*, who want to work for the certain level of wages, w_o, and the number actually employed, n', at this level of wages represented the size of involuntary unemployment.

In such conditions, according to Keynes, the initial assumption of a constant level of prices no longer applied because prices—expressed as units of wages—depended on the size of the wages and on the number of employed labourers. As a result, when the equilibrium level of employment, n', fell under the level of labour supply, n, at the inelastic level of wages, w_0, there would be a simultaneous decrease in prices or deflation.

In contrast, Keynes noted that an increase in demand above the level of the existing supply would partially result in an increase in prices and partially in an increase of employment and income, along with a decrease in real wages because of a decreasing marginal productivity of labour. This situation would prevail until the moment the system reached a state of full employment. At this moment, employing an additional unit of labour would require an increase in real wages, even as its real productivity decreased with a tendency towards zero. In such a situation, "any attempt to increase investment still further will set up a tendency in money prices to rise without limit, irrespective of the marginal propensity to consume; i.e. we shall have reached a state of true inflation," Keynes wrote.[20]

By synthesising the common determinants of the short-term movement of employment and prices and pointing out the disorders occurring in these areas when the equilibrium established or was in the process of establishing the general equilibrium in the system, Keynes defined the main goal of short-term macroeconomic policy as achieving full employment without "real" inflation. At the same time, he emphasised unemployment and inflation were the mutually related extreme consequences of the imperfection of the unique process of balancing supply and demand on the market. These consequences could be removed or mitigated only through organised activity of the economic policymakers.

Because, in Keynes' opinion, the main cause of these disturbances rested in deviations of the effective demand from the real level of supply at full employment, the main instruments of economic policy should be tax, fiscal, and credit-monetary policy measures which directly influenced the size, allocation, and use of income by harmonising the demand with a given supply. Keynes believed in the short term macroeconomic should be focused on ensuring a stable equilibrium in a system with the full employment of the variable production factors and without inflation.

The Functional Approach

Despite presenting a wide range of elements and implications, the theories and models brought up in this analysis—ranging from Smith and Marx to Marshall and Keynes—share a common feature. They each essentially highlight the fundamental problem of attempting to fit economic activity into the calendar time units on the basis of precise mathematical criteria, generally referring to the creation of partial equilibria, a general equilibrium, or both within a system. The problem becomes even more evident when analysing the totality of the causal, structural, and equilibrium (or disequilibrium) relations between the economic categories, as well as in the context of analysing the actual concept of time as a relative category.

From this perspective, not only quantitative changes occurring during economic activity, but also the qualitative-quantitative changes, are of significant importance.

Economic activity unfolds according to a certain pattern, and in a narrow sense this pattern is based on:

- Economic categories.
- Causal, structural, and equilibrium relations among these categories.
- General rules affecting the creation of these relations.
- Information systems.
- Dimensions of space and time.

In a wider sense, the pattern also comprises elements including:

- Structure of society.
- Motivational system.
- Organisation and methods of directing the economic activity.
- Natural conditions.
- Science and technology.
- Education.
- Ethics.

Elements included in the wider sense are indirectly manifested in this pattern or model of economic activity. Their indirect impact can be observed by their influence on the existence, form, content, efficiency, and the general behaviour of the main elements of this pattern. For instance, the influence of these factors on the sphere of the economic system and methods of directing the reproduction is primarily manifested in the number and character of the economic categories or variables in the system, overall system efficiency, structural relations in distribution and consumption, possible levels of detail and precision in forecasting the subsequent course of reproduction in the system, and other areas, while their influence on the sphere of applied science and technology is manifested in areas including the value of individual economic categories existing in the phase of production, structural relations between the factors of production, efficiency of the factors, and the values of the key transformation operators in the system.

Of course, this pattern is not fixed, but subject to larger or smaller changes in the historical continuity of the unfolding of the economic activity because of reasons external or internal to the functioning of the economic system. The intensity and character of these changes depend on which elements of the pattern change at a certain moment or period of time and reflect the level of stability for these elements.

Further, the changes in the pattern of functioning or reproduction of the economic system may refer to:

- Changes in numerical values of transformation operators and structural relationships, including their algebraic forms, among the economic categories in the system, that is, changes in the values and forms of parameters in the economic model.
- Changes in functional relationships among the sets of economic categories in the system, that is, changes in independent variables in relations or equations of the economic model.
- Changes in the content, character, and number of economic categories in the system, that is, changes in the very model of the economic activity.

In accordance with such changes in the pattern of economic activity, time periods of different duration may be classified as:

- *Very short term*: Economic activity takes place according to the features of a pattern that is stable in terms of all of its basic elements, considering the stochastic character of such patterns. In this period, all the relevant economic categories or variables and relations among these categories are unambiguous and fixed, while the empirical values of parameters do not vary significantly from the average and the parameters of the model.

- *Short term*: The pattern is stable in terms of determining the contents, the number and characteristics of economic categories in the system, and relationships among these categories, but the values of parameters could vary significantly from the average. In such a period, the changes in the functional modality of the system do not interfere with its categorical and institutional basis.
- *Medium term*: A restructuring period during which the relationships among the economic categories are changing along with the values of parameters, and in the process changing the circular flows and their reproductions and affecting the key qualitative and organisational features of the economic system.
- *Long term*: The categorical basis of the economic activity is also changing.
- *Very long term*: The overall organisation and the entire pattern of the economic activity are changing.

Relevant to this classification of time periods of different duration, two related elements are important to consider. First, expressing the temporal flow of the economic activity in units of the functional and relative "economic" time includes expressing the "time determination" of this activity in the same units of time. As a result, these periods do not necessarily have to be compatible with "short," "medium," or "long" periods of the calendar time. On the contrary, the modalities of the unfolding of the economic activity, which were mentioned among the characteristics of the short term, may exist in a period of several calendar years, while the changes in these modalities indicated as the characteristics of the medium term may in practice be realised in a relatively short calendar period.

And next, the economic activity and time "contained" in it flow continuously and this flow cannot be disaggregated into separate units independent of each other. Therefore, the important qualitative changes within the system do not occur outside of the time flows defined as short term. At the same time, because of the length of the transformation processes, they also cannot be treated as the main feature of the moment in which they were manifested in their final form.

These two elements emphasise the need to analyse short-term periods in the context of long-term trends and to view these trends as a function of the courses of reproduction in the shorter periods. However, their relation may not be stereotypically interpreted by dividing, for instance, the medium term into a certain number of short-term periods. Its character is strictly functional in terms of transformation of the quantitative or partial flows into the new qualitative feature of the system reproduction.

Under this analysis of the time flow and time determination of the economic activity, the traditional views on the objectives, tasks, and instruments of short- and long-term economic policy may also be modified to re-interpret their mutual permeation and abandon the idea of disaggregating short-term stabilisation problems from the long-term problem of economic growth. In essence, the analysis suggests policy should be not crafted by determining the stabilisation policy for harmonising the aggregated supply and demand without regard to the dynamics of the trends in production and also not by determining development policy on the assumption that rational long-run policy decisions automatically prevent any occurrence of significant disorders in the short-term flows.

Notes

1. François Quesnay, *Analize ekonomske tablice. Vol. 17–18, Ekonomisti* (Zagreb: Kultura, 1952).
2. Adam Smith, *An Inquiry into the Nature and Causes of the Wealth of Nations* (Hampshire: Harriman House, 2007) 37.
3. Ibid.
4. Ibid.
5. David Ricardo, *On the Principles of Political Economy and Taxation.* Chapter 4, *On Natural and Market Price* (Kitchener: Batoche Books, 2001) 54.
6. Alfred Marshall, *Principles of Economics.* Unabridged eighth ed. (New York: Cosimo Classics, 2009) 288.
7. Ibid., 305.

8. Ibid., 289.
9. Ibid., 285.
10. Ibid., 313.
11. Ibid., 314.
12. Charles E. Ferguson and John P. Gould, *Microeconomic Theory* (Georgetown: Irwin-Dorsey International, 1975) 126.
13. Edmund Malinvaud, *The Theory of Unemployment Reconsidered* (Hoboken: Wiley, 1977).
14. Malinvaud, *Unemployment Reconsidered*, 9.
15. Joost Van Doorn, *Disequilibrium Economics* (New York: Macmillan Studies in Economics, 1975) 76.
16. See Axel Leijonhufvud, *On Keynesian Economics and the Economics of J.M. Keynes* (New York: Oxford University Press, 1968).
17. John Maynard Keynes, *The General Theory of Employment, Interest and Money* (New Delhi: Atlantic Publishers and Distributors, 2008) 87–88.
18. Ibid., 121.
19. Ibid., 5.
20. Ibid., 107.

References

Doorn, Joost Van. 1975. *Disequilibrium Economics*. New York: Macmillan Studies in Economics.

Ferguson, Charles E., and John P. Gould. 1975. *Microeconomic Theory*. Georgetown: Irwin-Dorsey International.

Keynes, John Maynard. 2008. *The General Theory of Employment, Interest and Money*. New Delhi: Atlantic Publishers and Distributors.

Leijonhufvud, Axel. 1968. *On Keynesian Economics and the Economics of J.M. Keynes*. New York: Oxford University Press.

Malinvaud, Edmond. 1977. *The Theory of Unemployment Reconsidered*. Hoboken: Wiley.

Marshall, Alfred. 2009. *Principles of Economics*. Unabridged 8th ed. New York: Cosimo Classics.

Quesnay, François. 1952. *Analize ekonomske tablice. Vol. 17–18, Ekonomisti*. Zagreb: Kultura.

Ricardo, David. 1817. *On the Principles of Political Economy and Taxation*. Kitchener: Batoche Books.

Schumpeter, Joseph. 1954. *The History of Economic Analysis*. New York: Oxford University Press.

Smith, Adam. 2007. *An Inquiry into the Nature and Causes of the Wealth of Nations*. Hampshire: Harriman House.

14

Concluding Remarks

The analysis offered on the nature, character, and content of the time dimension of economic activity and its impact on the modalities of economic activity leads to a series of conclusions.

(1)

Traditional analyses of the time dimension of economic activity are usually based on the perspective of classical physics, which presents the universal and absolute character of time, unaffected by events and processes occurring within time. Within such an approach, the values of economic categories and flows of their reproduction were determined by using the methods of aggregation within elementary calendar time units. This created a concept about the type, character, and modalities of establishment and numerical values of the causal, structural, and equilibrium relationships among economic categories in the selected unit of calendar time, while also acquiring the knowledge about the state of the economic activity in this time unit or about the change of the state within this unit.

This process was primarily based on objective factors, such as the complexity of the economic activity, a large number of economic subjects, limitations within information systems, needs to simplify the analyses,

Ž. Rohatinski, *Time and Economics*, DOI 10.1007/978-3-319-61705-3_14

and others. However, this approach also conceals the threat of "distortion" of actual modalities of economic activity, especially in establishing the forms and values of the transformation operators and the equilibrium relationships among economic categories in the system.

(2)

To avoid such distortion, economic activity should be viewed using a different concept of the time dimension based on a series of observations about the nature and character of time:

- Absolute time passing evenly from the point of chronology is merely an abstraction in which the objective functional times of different processes overlap.
- Each individual process exists in an internal, functional time that is specific and independent from any possible changes in the external time measurement system.
- The internal time measurement depends on the objective properties of a certain process, while the integration of time flow and its measuring system makes the process temporal.
- The subjective cognition of time represents a static awareness of the sequence of the processes and their duration, which becomes dynamic because of the existence of an observer in the objective time.

(3)

To understand the implications of these observations of the time dimension of economic activity, the main features, character, content, and modality of economic activity must be clearly defined.

(3.1) Economic activity is a significantly complex, multi-level activity that enables individual subjects and the society to organise, implement, and direct the material and social processes aimed at ensuring the basic precondition of their existence and development.

(3.2) The main entity of economic activity is an individual economic event creating a relationship among two or more economic subjects or among these subjects and nature. Divisible and structurally heterogeneous in technical, legal, or sociological sense, they are unique and homogenous events from an economics perspective.

(3.3) A functionally rounded group of economic events of the certain content, character, and type of activity, such as production, distribution, exchange, and consumption, interrelated with the functional relations of the sequence, mutual influence, and aggregation represents a process of the certain level of complexity, frequency, and modality of occurrence.

(3.4) Functional merging or concatenation of different types of economic processes creates a circular flow of economic reproduction of subjects taking part in these events and processes, while their functional unity represents the economic activity.

(3.5) Society has numerous natural, technical, financial, and other types of resources that form the basis for pursuing economic activity. When included into economic events and processes, these resources become the corresponding economic categories. Consequently, a complex group of economic relationships is created among these categories:

- *Causal relationships*: An essential characteristic of any economic category, determine the main mechanism of establishing functional relations among the economic events and processes by way of qualitative-quantitative transformations of economic categories into a unique course of the economic activity of the certain subjects and a society.
- *Structural relationships*: Established between functionally related economic categories during an economic event or in the same phase of a circular flow of economic activity, represent a basis of homogenisation or diversification of individual circular flows of reproduction of economic categories.
- *Equilibrium relationships*: A regulated set of causal and structural connections among economic categories, ensuring regular reproduction and economic activity

(3.6) Continual repetition and parallel activity of circular flows of reproduction of economic categories in the course of economic activity of individual subjects and a society as a whole generate the common qualitative and quantitative elements. They are present within certain phases of these flows as an average or a tendency and enter the structure of almost all equivalent causal, structural, and equilibrium relationships among relevant economic categories. Manifested at the level of the economic activity of a society, these elements represent the general rules of such an activity with a stochastic, historical, and objective character.

(3.7) A socially comprehensible set of categories crucial for economic activity, interrelated with the common ratios of the causal, structural, and equilibrium character, represent a model of such an activity.

(4)

Within this framework, the analysis considers the functional time flow or the "economic" time of economic activity in light of a modified version of Marx's model of reproduction of an individual money capital. Several points deserve emphasis.

(4.1) According to classical political economics and the labour theory of value, the time course of reproduction of individual money capital includes a labour period, a period of excess production time over labour time, and a period of trade, which together form a period of a single circular flow of money capital reproduction. The duration of such periods is defined by the calendar or chronological time of the duration of economic processes that allow capital to enable an individual product in a production process to acquire its use value and exchange value and to enable the transformation of money capital into commodity capital and commodity capital into money capital in the process of trade. The sum of the durations of these processes represents the duration of the period of a single circular flow of the reproduction of money capital.

Since such a period includes several different processes—labour, production, and trade, for example—the period is not homogenous in terms of content, time, or space. However, the functional aggregation of economic processes in economic activity of the specific subject leads to homogenisation of the content, time, and space of the economic processes representing the economic activities of this subject in relation to the content, time, and space of the economic activity of other subjects.

In this sense, the duration of a labour period is essential for the time flow of economic activity and the related relationships between the economic categories established in a single circular flow of the money capital reproduction. Other periods or subsets within this period might be defined as functions of the duration of a labour period.

On this basis, the turnover time of the total money capital paid in advance, with its calendar length determined by the duration of the instruments of labour, might be defined as a result of the duration of the period of a single circular flow of the money capital reproduction and a number of products that were produced discretely and repeatedly and realised in the course of this turnover. Since the duration of the circular flow of the money capital reproduction may be expressed as a function of the duration of a labour period, the turnover time of the total money capital paid in advance may also be expressed in the units of the duration of this period.

(4.2) Because the duration of the labour periods in reproduction of individual money capital paid in advance is generally different and varies in different stages of development of individual capitals and of the total capital in a society, each individual capital in the certain stage of development has a functional measuring unit of its turnover time. Further, the duration of its reproduction time is different. Such reproduction time, expressed in the units of a labour period to which this capital is subjected, may be called the functional reproduction time.

The difference in the duration of labour periods as a measuring unit of the turnover time of individual money capitals paid in advance and the resulting differences in functional reproduction times of such capital imply that:

- Equal calendar durations of the turnover time of two or more individual money capitals paid in advance are not functionally equal if measured in units of their labour periods.
- In cases where the calendar duration of their periods of excess production over labour period is equal to the periods of trade or the difference in the sum of their durations does not match the inverse difference in the duration of the labour period, then the calendar duration of the periods of a single circulation flow of the capital reproduction also cannot be equal.

With the reproduction of an individual money capital paid in advance, the same calendar time unit may have different functional "densities" because in the course of equal calendar durations of turnover of two or more individual money capitals paid in advance, a different quantity of circular flows of their reproduction may be realised.

(4.3) An individual capital unit cannot be reproduced autonomously since in the phase of exchange it is bound to establish the value and purpose relationships with other capital units. Therefore, for normal capital reproduction and to establish equilibrium in a system, at the moment of exchange these capital units should be in the appropriate form, such as cash or commodity, as well as the appropriate size. Taking into account the different functional time flows of the individual capital reproduction, this requirement necessitates a specific constellation of causal and structural relationships among economic categories in the system, which, by comprising a functional difference of the time attributes of the origin, existence, and reproduction of the certain categories, will result in a simultaneous establishment of both content-related and time-related equilibrium of the system.

(5)

Analysing these perspectives using a model of the economic activity that includes consideration of the relativity of time produces insights into the nature, character, and role of the time component of economic activity. For this, the common capital reproduction model must be modified allowing all variables to be expressed in units of labour periods in an individual sector. The analysis leads to several significant conclusions about the nature, character, and role of the time component of economic activity.

(5.1) Each economic system, as a subset of the macroeconomic system, is reproduced in its internal or functional time, while the base and a measuring unit of this time is the length of a labour period within this system. This measuring unit is determined by the set of the natural, technical, and economic conditions, representing the objective or relatively objective factors for the given economic system in the specific stage of development. All of the time features of the reproduction flow in the system, such as the period of a single circular flow of the capital reproduction, the period for which the wages are paid, and duration of the initial advance instruments of labour, expressed in the units of the duration of a labour period represent the functional time flow of the system reproduction or the functional time of the system. The functional time flow is inherent to the origin, existence, and termination in the certain manifestation of any economic category or variable in the system and implicitly built into all causal and structural relationships parameters between those categories.

(5.2) The duration of the certain phase or process in the course of economic activity of a system is determined by the sum of the consecutive functional time units, in which the entire time-space pattern of this process is realised. This length is "absolute" from the aspect of the time measuring system in the specific system, but relative in terms of relationships between the time measuring systems in the systems moving at a uniform speed. This means that in the system moving relatively to a stationary system, a time dilation can be observed. Dilation may be analytically described by applying a modified form of Lorentz transformations, where the transformation factor represents the function of speed of the system, turnover time of the total capitals paid in advance in the systems, and the length of duration of the labour periods in an individual system. This is also related to the issue of different efficiency of the systems in the same "non-temporal" conditions of reproduction.

(5.3) Normal reproduction of two or more functionally related economic systems in different functional times requires establishing an integral material-financial-time balance of the macroeconomic system. In other words, the relationship between the functional reproduction time flows of individual systems directly affects the allocation and structure of the use of capital in the macroeconomic system, the equilibrium size of the certain economic categories or variables in each system, structural and causal relationships or parameters among these categories in an individual system, absolute and relative efficiency of each individual system, and the time modalities of establishing a macroeconomic balance. This means that the functional times of individual systems are the active factors that initially and recurrently affect the total forms of reproduction in these systems and in the entire macroeconomic system.

(5.4) The functional unity of functional times of every system within the macroeconomy represents the functional time of the macroeconomic system. Since the macroeconomic system consists of the numerous systems that differ in size and complexity, in hierarchical levels, and in activities that build upon, overlap, and oppose one another, the functional time of the macroeconomic system is difficult to view empirically. Therefore, in practice, the time flow of its economic activity is mostly expressed, analysed, and measured in the calendar time units.

However, from the aspect of economic activity of an individual system and the macroeconomic system as a whole, the calendar time units are generally non-functional. As a result, this compulsory temporal aggregation of these activities into uniform calendar time units results in losing insight into their actual or functional flow and creating a partially distorted image of the efficiency of such activities. Three related factor drive this:

- Because of different durations of labour periods, periods of circular flow of capital reproduction, and the turnover time of the total advanced capital in the individual systems, the same calendar time unit contains different amounts of the functional time units of an individual system and, in turn, a different number of economic events and processes representing a functional part of one or more circular flows of reproduction of the total advanced capital paid in an individual system. At the same time, these calendar time units have different densities, as well as variances in their functional scope and in their content. Unambiguous aggregation of such different contents in a fixed calendar time unit prevents a comprehensive view of the functional flow of the system reproduction within the calendar time unit.
- Aggregation into calendar units sums up the values of the variables within the individual systems and in the macroeconomic system, which functionally pertain to different circular flows of capital reproduction and turnovers of the total advanced capital. This distorts the actual value of parameters among the variables within the system, which is functionally established in the unit of the circular flow of reproduction.
- The calendar time units transect this flow in an individual system in different ways and in different functional phases of the flow of reproduction. This distorts the perception of actual functional relationships between the systems and also the perception of the modality of functioning of the macroeconomic system.

Expressing the temporal dimension of the economic activity in the calendar time units may only simulate the actual time flow of the economic activity of an individual economic subject or system and the society as a whole to the limited extent, but cannot completely substitute for the functional utility of comprehending this flow as functional time of each individual system and of the macroeconomic system.

(6)

The characteristics, nature, and character of this perspective of "economic" time result in different definitions of the notion of "maturity" of economic activity and classifications of time periods of different functional durations in relation compared with the approaches followed by Smith, Marshall, and Keynes. The duration depends on the characteristics and modalities of the economic activity of individual subjects and the society in each of these periods. The criteria of this classification represent a "functioning model" or "pattern," where the intensity and the character of changes depend on the elements of the pattern changing in the specific moment or period.

In this light, the changes in the pattern of functioning of the economic system may refer to:

- Changes in numerical values of transformation operators and structural relationships, along with their algebraic forms, among the economic categories in the system, that is, changes in the values and forms of parameters in the economic model.
- Changes in functional relationships among the sets of economic categories in the system, that is, changes in independent variables in relations or equations of the economic model.
- Changes in the contents, character, and number of economic categories in the system, that is, changes in the very model of the economic activity, including changes to the entire economic system.

By systematising changes in the pattern of economic activity, the time periods of different duration may be classified as:

- *Very short term*: Economic activity takes place according to the features of a pattern that is stable in all of its basic elements, considering the stochastic character of such patterns. In this period all the relevant economic categories or variables and relations among these categories are unambiguous and fixed and the empirical values of parameters or the model parameters do not vary significantly from the average.

- *Short term*: The pattern is stable in terms of determining the content, the number and characteristics of economic categories in the system, and relationships among these categories, but the values of parameters could vary significantly from the average. This means that in this period, the changes in the functional modality of the system do not interfere with its categorical and institutional basis.
- *Medium term*: The relationships among the economic categories are changing along with the values of parameters, and as a result the circular flows and their reproductions are also changing, affecting key qualitative and organisational features of the economic system. This can also be seen as a restructuring period.
- *Long term*: The period when the categorical basis of the economic activity is also changing.
- *Very long term*: The overall organisation and the entire pattern of the economic activity are changing.

(7)

The time dimension of the economic activity comprises the functional time flow of economic activity of individual subjects and society, together with the time determination of that activity, which depends on the characteristics of the pattern according to which this activity unfolds in the functional time.

(8)

Depending on the characteristics and attributes of the time dimension, the forms, the manner, and the intensity of directing the flows of reproduction of economic categories taken by a society to ensure the normal reproduction also change. In the very short term, during which the pattern of economic activity is stable in all of its elements, economic policy has no significant impact on the existing "quality" of the functioning of the economic system and its potential "technical" efficiency. However, by influencing the size of the engaged production factors and the possibility of product realisation, policy affects the quantitative level of system reproduction and a level of engagement and utilisation of the production factors. In contrast, in longer periods of functional time, economic policy might affect the form and value of the transformation operators, formation of functional relationships among the economic categories in a system, and the structure and contents of these categories. This affects the qualitative, and through them also the important quantitative characteristics of the functioning of the economic system.

* * *

Beyond the genealogy and formulas, what do these theses on functional time in economic activity actually tell us? How do they expand our views of classical political economics?

They certainly talk of the need to consciously include the time component in any analysis of economic processes at the micro- and macroeconomic levels. But, they are saying much more.

These concepts show that the modalities of the unfolding of economic activity and the reasons behind the concrete actions of its main entities are much more complex and sensitive than is often presented in theories and models based on "market fundamentalism." Numerous contemporary models are built around hypotheses that suggest a perfect awareness of all actors in the market, prices that serve as the key channel for information, maximisation of profit as the primary driver of economic activity, "unlimited" supply and demand on the global market of production factors, commodities, and services, a high level of substitutability of products, labour, and capital, and high supply and demand elasticity. Under these conditions, economic processes described by these models seem rational, readily understandable, and predictable, whether describing the past, explaining the present, or predicting the future.

Unfortunately, the real world outside these models is not so simple. For cohesion, these models tended to abstract many relevant economic and non-economic factors that affect the flow of economic activity or to treat these factors as self-explained, residual influences. Among these neglected factors, time stands out as critical to a complete understanding of how economic activity unfolds. As the analysis showed, time flows differently in individual segments of economic activity and, just as important, shapes those segments differently. Even if the social character of economic activity makes a precise mathematical definition of its influence elusive, a conceptual understanding of the time element of economic activity would aid public leaders as they work towards more efficient policies and economists as they develop more nuanced models.

Appendix A: Transformation Problem

The form and value of the transformation operators and the modalities of their formation are among the key research areas in economics. Within this framework, a prominent position in the history of economic analysis belongs to the issue of transformation of the value of commodity into its price, which is generally known as the "transformation problem." It was explicitly raised first by Marx in Chap. 10 of the Volume III of *The Capital* and later analysed by a number of authors with different theoretical and conceptual views, Ladislaus Bortkiewicz, Joseph Winternitz, Francis Seaton, Paul Sweezy, William Baumol, Paul Samuelson, Michio Morishima, and others. The essence of this problem is reduced to the issue of the method of equalisation of the profit rates at different organic compositions of individual capitals and in accordance with that, the formation of prices of commodities provided that their level and relative relations are sufficient to satisfy three conditions:

- The product of the price and quantity of all products equal their value determined according to the labour theory of value.
- The total profit equals the total surplus value.
- The process of reproduction is normal, meaning that all the conditions of equilibrium are fulfilled.

Ž. Rohatinski, *Time and Economics*, DOI 10.1007/978-3-319-61705-3

From this, Marx first concluded that in different branches of industry there are different profit rates even in case of the same rates of surplus value. This was in line with different organic composition of capitals and times of their turnover. In his opinion, this was incompatible with the main assumption of the functioning of the capitalist system and caused continuous relocation of money capital from the productions of lower profit rate to the productions of higher profit rate. Parallel to growing competition in the more profitable productions and reduced competition in less profitable productions, this led to a tendency of equalisation of profit rates in different productions and a tendency of their equalisation with an average profit rate. In such circumstances, the "cost price" of the product, which corresponds to the value of the employed constant and variable capital, is transformed into the price of production representing the cost price increased by the profit resulting from the average profit rate applied to the size of the individually employed constant and variable capital.

Marx illustrated this modality of transforming the value of products into their price by using an example of five producers with different capital compositions, but the essence of his example may also be presented by using the standard two-sector model and the scheme of reproduction in two sectors.

On the basis of the original Marx's numerical example, the values and relations in the value structure of the products in the first and second sector, the means of production and articles of consumption, are illustrated in Table A1.

Application of an average profit rate, 24 percent, to such production values of the first and second sector, leads to the formation of production prices for their products shown in Table A2.

Table A1 Transformation of the value of product into the price of production - part 1

Sector	Employed capital: Constant (1)	Employed capital: Variable (2)	Surplus value (3)	Value of production (4)	Organic composition of capital (5) = (1)/(2)	Profit rate $(6)=\frac{(3)}{(1)+(2)}$
I	4000 P	+ 1000 P_r	+ 1000 V	= 6000 W	4	20%
II	1500 P	+ 750 P_r	+ 750 V	= 3000 W	2	33%
Total	5500 P	+ 1750 P_r	+ 1750 V	= 9000 W	3.2	24%

Source: Author's calculations

Note: The scheme is provided that the surplus value rate is 100% and equal in both sectors

Table A2 Transformation of the value of product into the price of production - part 2

Sector	Cost price (1)+(2) (7)	Profit 0.24/(1)+(2)/(8)	Production price (9)
I	5000 C_k	+ 1207 Prof.	= 6207 C_p
II	2250 C_k	+ 543 Prof.	= 2793 C_p
Total	7250 C_k	+ 1750 Prof.	= 9000 C_p

Source: Author's calculations

If we compare the results obtained by transforming the value into the price of production, it is evident that they meet the requirement of equality between the price of production of all products and their price, 9000 money units, as well as the requirement of equality between the total profit and the total surplus value. This indicates that the transformation was carried out on the basis of redistribution of the surplus value created in the second sector, which was, by means of the price of production, transferred to the first sector. However, these results do not fulfil the requirement of the system equilibrium, which was used by many critics of Marx's work as an argument to try to refute the entire labour theory of value. Namely, the general equilibrium of the economic system in the circumstances of expanded reproduction is defined by the following relations:

$$P(I)+\Delta P(I)+P(II)+\Delta P(II)=P(I)+P_r(I)+V(I)$$

$$P_r(I)+\Delta P_r(I)+V(I)+P_r(II)+\Delta P_r(II)+ V(II)=P(II)+P_r(II)+V(II)$$

$$P_r(I)+\Delta P_r(I)+V(I)=P(II)+\Delta P(II)$$

Starting from Marx's assumptions that, for the purposes of expanded reproduction, the first sector would advance 50 percent of achieved profit for employment of the production factors in a ratio resulting from the given organic composition of capital (4), the profit of the first sector would be used for the following purposes:

- 604 money units for the private consumption of the capitalist.
- 483 money units for purchasing the means of production or constant capital.
- 120 money units for acquiring additional labour power or variable capital.

In such conditions, the demand of the first sector for the products of the second sector would rise by 120 money units, which must be accumulated from the profit of the second sector to acquire additional means of production in order to maintain the equilibrium in the system. At the given organic composition of capital (2), the acquisition of the additional means of production required that the second sector employ additional labour power in the value of 60 money units for their total accumulation to amount to 180 money units, while 363 money units of the profit would be retained for private consumption of the capitalists of the second sector. This would lead to the formation of the following relations between supply and demand for products of the first and second sector:

$$4000P(I)+483\Delta P(I)+1500P(II)+120\Delta P(II)=6103$$
$$1000P_r(I)+120\Delta P_r(I)+604V(I)+750P_r(II)+$$
$$604P_r(II)+363V(II)=2897$$

This result means that the total demand for products of the first sector was 104 money units lower than the production of this sector expressed in the units of production (6207–6103). This is also the difference by which the total demand for the means of consumption exceeds the production of the second sector, expressed in the prices of production (2897–2793). This indicates that transferring part of the surplus value from the first to the second sector, in accordance with Marx's solution of the transformation problem, would result with a situation in which, according to the third condition of the system equilibrium:

$$1000P_r(I)+120\Delta P_r(I)+604V(I)>1500P(II)+120\Delta P(II)$$

Demand of the first sector for the products of the second sector would exceed by 104 money units the demand of the second sector for the products of the first sector, resulting in a state of disequilibrium.

The same conclusion for a system equilibrium disorder when transforming product value into the price of production in the manner assumed by Marx was reached on the basis of the corresponding examples referring to simple reproduction. This leads to the conclusion that the value of the transformation operator does not represent merely a weighted average of the profit rates that would be, depending on their organic composition, achieved by the individually employed capitals.

In his attempt to resolve this problem, Bortkiewicz claimed that Marx erred when he assumed that the production factors were acquired at the price of their value and not at the price of their production, which already has a built-in profit as the transformed surplus value of the producer of the articles of consumption bought by capitalists and of the producer of the articles of consumption bought by labourers. Because of this, when transforming the value of the new product into his price, Marx left the value of the production factors to remain untransformed. To avoid this "mistake" and to determine the value of the transformation operator which would fulfil the conditions of the system equilibrium, Bortkiewicz created a three-sector scheme of the simple reproduction:

- Production of the means of production.
- Production of the articles of consumption of labourers.
- Production of luxury articles of consumption for capitalists.

This scheme implicitly contained the conditions of the system equilibrium:

$$P(I)+P_r(I)+S(I)=P(I)+P(II)+P(III)$$
$$P(II)+P_r(II)+S(II)=P_r(I)+P_r(II)+P_r(III)$$
$$P(III)+P_r(III)+S(III)=S(I)+S(II)+S(III)$$

Note: S = profit.

Bortkiewicz continued by introducing as unknowns average purchase prices for the means of production (x), articles of consumption for labourers (y), and articles of consumption for capitalists (z), as well as an average profit rate (r).

From this follows that the relations determining the conditions of the system equilibrium became a set of three equations with four unknowns:

$$xP(I)+yP_r(I)+r\big(xP(I)+yP_r(I)\big)=x\big(P(I)+P(II)+P(III)\big)$$
$$xP(II)+yP_r(II)+r\big(xP(II)+yP_r(II)\big)=y\big(P_r(I)+P_r(II)+P_r(III)\big)$$
$$xP(III)+yP_r(III)+r\big(xP(III)+yP_r(III)\big)=z\big(S(I)+S(II)+S(III)\big)$$

Expressed otherwise:

$$(1+r)\big(xP(I)+yP_r(I)\big)=x\big(P(I)+P(II)+P(III)\big)$$
$$(1+r)\big(xP(II)+yP_r(II)\big)=y\big(P_r(I)+P_r(II)+P_r(III)\big)$$
$$(1+r)\big(xP(III)+yP_r(III)\big)=z\big(S(I)+S(II)+S(III)\big)$$

To solve this set of equations, Bortkiewicz further assumed that gold, as the general equivalent of the product value in the section of producing the luxury articles of consumption, was represented by $Z = 1$, which reduces this system to three equations with three unknowns. After substitution for the purposes of simplified presentation of the expression:

$P_r(I)/P(I)$, with f_1,
$P_r(II)/P(II)$, with f_2,
$P_r(III)/P(III)$, with f_3,
$\big(P(I)+P_r(I)+S(I)\big)/P(I)$, with g_1,
$\big(P(II)+P_r(II)+S(II)\big)/P(II)$, with g_2,
$\big(P(III)+P_r(III)+S(III)\big)/P(III)$, with g_3
$(1+r)$, with m

The solution to the equation is:

$$m=\frac{f_2g_1+g_2-\sqrt{(g_2f_2g_1)^2+4f_1g_1g_2}}{2(f_2-f_1)}$$

$$r = 1 - m$$

$$y = \frac{g_3}{g_2 + (f_3 - f_2)m}$$

$$i \cdot x = \frac{f_1 ym}{g_1 - m}$$

From this follows that the value of the transformation operator in the transformation of the production value of the sector i ($i = 1, 2, 3$) into the production price of this sector, presented as a multiplier, is expressed as:

$$T(i) = \frac{xP(i) + yP_r(i) + r(xP(i) + yP_r(i))}{W(i)}$$

Because of the initial analytical assumptions—the model is based on the relations determining the system equilibrium—Bortkiewicz's solution to the transformation problem a priori satisfies the condition of equilibrium. However, the conditions of equality of the production price and value and equality of the profit and surplus value in the overall economy may be fulfilled by this solution only under the assumption that the organic composition of capital in production of gold, as the general equivalent, equalled an average composition of capital of the overall economy, which can only be a random event. In contrast, this procedure leads directly to a difference between the sum of the production price of all commodities in a society and the sum of their value. In addition:

- The solution is valid in a specific case only for simple reproduction in a three-sector economy structure.
- There is an unresolved issue of individual productions, which should have the same organic composition of capital, aggregated into sectors.
- A derivative that shows the profit rate does not depend on surplus value created in production of the means of production and in production of the articles of consumption for labourers. (This also applies to the influence of the organic composition of capital on the formation of the profit rate.)
- Other weaknesses exist.

These aspects impose further restrictions to a degree of generality applicable to Bortkiewicz's solution.

Despite these weaknesses, in a general analysis of the transformation operator this solution is relevant for introducing a transformation operator directly into the procedure. This operator facilitates the transformation of value into the price of commodities and the price of input in production of these commodities, which are already transformed in relation to the value of these inputs. Bortkiewicz clearly showed that the value of an individual transformation operator does not depend only on the value and relations between the value of elements existing in the phase of the economic activity referenced by the respective transformation operator, but also on the totality of relations among all the elements of the economic activity of the individual subject in a society. In a certain phase of economic activity, depending on the specifics of this phase, these relations create a specific constellation.

Such an approach exists in all subsequent analysis of the transformation problem, even as:

- Some authors, including Winternitz, produce only modifications of the certain elements of Bortkiewicz's model.
- Authors, including Seton, explicitly criticise this model by searching for new solutions.
- Authors, including Morishima, analyse Marx's approach by applying modern mathematical methods and to prove that the essence of his solution of the transformation problem was correct, in other words that the transformation of the value of input into their prices is carried out successively through an adequate iterative procedure.

For more detail, please see:

Meek, Ronald. *Smith, Marx & After*. Belgrade: Serbian Research-Publishing Centre, 1981, pp. 122–140.

Morishima, Michio and George Catephores. *Value, Exploitation and Growth*. Zagreb: Centar za kulturnu djelatnost, 1985, pp. 131–154.

Seton, Francis. The Transformation Problem. *Review of Economic Studies*, No. 24. 1957, pp. 149–160.

Sweezy, Paul M. *The Theory of Capitalist Development*. Zagreb: Naprijed, 1959, pp. 122–143.

Appendix B: Conditions for Moderate Growth in a Two-Sector Economy

Eight relations are needed to consider conditions for moderate growth in a two-sector system.

(1)

The increment in the production value in the first and second sector is a function of size in the previous circuit of the capital of formed accumulation, which means that the production growth rate is a function of the share of accumulation in the value of production of the individual sector in the previous circuit of capital.

(2)

The absolute size of accumulation in the first circuit of capital in the first sector depends on the size of the capital advanced to this sector, relative size of capacity, rate of depreciation, organic composition of capital, rate of exploitation, and rate of savings or accumulation. With the exception of the rate of accumulation, these elements are assumed to be relatively objectively ex ante determined by the level of development of the means of production, technical and technological characteristics of the production process, historical price of labour power, and other factors.

In contrast, the rate of accumulation is relatively autonomously determined by the holders of the reproduction of capital depending on

Ž. Rohatinski, *Time and Economics*, DOI 10.1007/978-3-319-61705-3

the trend of demand for products of the first sector, impact of decisions made by the higher hierarchical structures, and their personal preferences and interests. If there are no changes of ex ante determined elements in the observed circular flow of reproduction in all circuits of capital of the first sector in which the rate of accumulation is fixed, in such a case the share of the accumulation in the value of production will be fixed, which means that the rate of the production growth will also be fixed. Consequently, the instruments of labour advanced to the first sector will grow at the fixed rate that is equal to the rate of the production growth. From this, the share of the "cumulative" disequilibrium between the supply of products of the first sector and demand of this sector for these products in the value of production of the first sector will be fixed.

(3)

The accumulation in the second sector is essentially defined by the same elements as the accumulation in the first sector, with one important difference. The rate of accumulation, according to Marx's schemes of reproduction, is not relatively autonomously defined by the capital reproduction holders of the second sector, but is directly affected by the size of the share of accumulation to be advanced by the first sector to employ additional labour power, as well as by the organic composition of capital of the second sector.

(4)

Assuming that the rates of depreciation and exploitation and the relative sizes of capacity in the first and second sector are equal, the ratio between the size of accumulation of the first and second sector in an individual circuit of capital will depend on the ratios between the size of production, as a variable element, and the organic compositions of capital, as a fixed element, of the first and second sector. Considering that in his numerical example Marx assumed that the value of production of the first sector in the first circuit of capital is twice as high as the value of production of the second sector and that the organic composition of capital in the sectors is also in that relation, it is evident that the ratio between the size of accumulation of the first and the second sector in the first circuit of capital will significantly exceed the ratios between the sizes of production in these sectors. This will result in a significantly lower share of the size of accumulation in the structure of the value of production of the second sector, in the first circuit of capital, compared with the share of the size of accumulation in the structure of the value of production of the first sector. As a consequence, the rate of production growth of the second sector in the second circuit of capital would be significantly lower than the rate of production growth of the first sector. Therefore, the subsequent flow of reproduction would result in an increased disequilibrium between the supply of products of the first sector and demand for these products in the second sector, resulting in an increase of the relative total cumulative disequilibrium between supply and demand for the products of the first sector.

(5)

The higher rate of production growth of the first sector compared with the rate of production growth of the second sector in the subsequent circuit of capital would also lead to an increase in ratio of the value of production of the first and of the second sector. If the share of accumulation in the structure of the value of production of the first sector were fixed, this would also result in an increase of the ratio between the size of accumulation of the first sector and the value of production of the second sector, which would create the need to increase the share of own accumulation of the second sector in the structure of the value of production to satisfy the increased demand of the first sector for the products of the second sector. Thus, despite of the increased ratio between the value of production of the first and second sector in the subsequent circuit of capital, in this circuit the ratio between the size of accumulation of the first and second sector is decreased, while the share of accumulation in the structure of the value of production of the second sector is increased.

(6)

An increase in the share of accumulation in the structure of the value of production of the second sector leads to an increased rate of production growth of the second sector, thus reducing the ratio between the fixed rate of production growth of the first and of the second sector. However, as long as the rate of production growth of the second sector is lower than the rate of production growth of the first sector, the ratio between the value of production of the first and of the second sector continues to grow, which results in the further increase of the ratio between the accumulation of the first sector and the value of production of the second sector, reflecting the increase of the share of accumulation of the second sector in the value of production of the second sector. This creates the prerequisites for an accelerated production of the second sector and the further reduction of the ratio between the rate of production growth of the first and of the second sector.

Here, it is significant that the accelerated rate of production growth of the second sector and a decreased ratio between the rate of production growth of the first and the second sector would result in an increase in the ratio between the value of production of the first and of the second sector. As a consequence, the intensity of change in all other analysed relations would be reduced, which would result in the intensity of acceleration of the rate of production growth of the second sector in the subsequent circuit of capital being lower than that reached in the previous circuit.

(7)

In subsequent circuits of capital these tendencies continue, whereby the rate of production growth of the second sector, moving along the hyperbolic curve, tends to become equal with the rate of production growth of the first sector when the changes in the ratio between the value of production of the first and of the second sector tend towards zero. Accordingly, in the certain circuit of capital (n) the rate of production growth of the second sector will be equal to the rate of production growth of the first sector when in the previous circuit of capital ($n-1$) the ratio between the value of production of the first and of the second sector, as well as other analysed relations, reaches certain fixed values. From then on, the rate of production growth of the first and the second sector remains unchanged for all subsequent circuits of capital and the growth of production in the system becomes moderated, while the relations between accumulations and productions of the first and of the second sector become fixed.

(8)

By equalising the rates of growth in the value of production of the first and of the second sector, the rates of growth in demand of the second sector for the products of the first sector and of supply of products of the first sector are also becoming equal, thus causing the relation between the absolute values of these categories to become fixed. Consequently, the "cumulative" share of unrealised products of the first sector in the value of production of the first sector also becomes fixed.

To present these relations mathematically, a series of definitions is needed (numbered for reference):

$$A(I) = \Delta SR(I) + \Delta PR(I) + \Delta SR(I) \quad (1)$$

$$w(I) = (a'\Delta SR + \Delta PR)(I) : \Delta RS(I) \quad (2)$$

$$\Delta SR(I) = \frac{1}{\mathrm{k}}\Delta PR(I) \quad (3)$$

$$\Delta RS(I) = \frac{1 + \frac{a'}{k}}{\mathrm{w}}\Delta PR(I) \quad (4)$$

$$RS(I) + \Delta RS(I) + A(I) - a'SR(II) - PR(II) = \Delta SR(II) + \Delta PR(II) \quad (5)$$

$$\Delta SR(II) = \frac{1}{\mathrm{k}}\Delta PR(II) \quad (6)$$

$$w(II) = (a'\Delta SR + \Delta PR)(II) : \Delta RS(II) \quad (7)$$

$$A(II) = \Delta SR(II) + \Delta PR(II) + \Delta RS(II) \quad (8)$$

Note: SR = instruments of labour, PR = objects of labour, RS = labour power, A = accumulation, k = relative size of capacity, w = organic composition of capital, a' = rate of depreciation, s = rate of accumulation or savings, se = rate of surplus value, and (I), (II) = the first and second sector.

Designating the value of production of the first sector in the main circular flow of capital reproduction as $W_o(I)$, the value of production in the subsequent circuit of capital is defined as:

$$W_1(I) = W_o(I) + a'\Delta SR_1(I) + \Delta PR_1(I) + \Delta RS_1(I) + se\Delta RS_1(I) \quad (9)$$

From here, according to expression (1), the growth in the value of advanced production factors equals the accumulation achieved in the previous circuit of capital:

$$\Delta SR_1(I) + \Delta PR_1(I) + \Delta RS_1(I) = A_o(I) \tag{10}$$

The size of accumulation results from:

$$A_o(I) = sV_o(I) \tag{11}$$

While the surplus value is determined by:

$$V(I)_o = \frac{seW_o(I)}{w(I) + 1 + se} \tag{12}$$

From here, the size of accumulation may be expressed as:

$$A_o(I) = \frac{W_o(I)}{w(I) + 1 + se} \tag{13}$$

Including expression (13) in expression (10) results in:

$$\Delta SR_1(I) + \Delta PR_1(I) + \Delta RS_1(I) = s\frac{W_o(I)}{w(I) + 1 + se} \tag{14}$$

And with this, the growth of any factor may be expressed as a function of the value of production in the previous circuit.

In this context it is necessary to determine the relations between the rates of growth in the individual production factors, and, therefore, to express the left side of expression (14) as the growth in a single production factor.

Starting from the expression (3), the growth in the value of the objects of labour is:

$$\Delta PR_1(I) = k\Delta SR_1(I) \tag{15}$$

And, starting from expression (2) the growth in the value of labour power is:

$$\Delta RS(I) = \Delta SR_1(I)\frac{a'+k}{w(I)} \tag{16}$$

Including the expressions (15) and (16) in expression (14) results in:

$$\Delta SR_1(I) = \frac{sW_o(I)}{\left(w(I)+1+se\right)\left(1+k+\dfrac{a'+k}{w(I)}\right)} \tag{17}$$

From this it follows that by including expression (17) into expressions (15) and (16), and further by including all three expressions into expression (9), the value of production of the first sector in the subsequent circuit of capital is:

$$W_1(I) = W_o(I) + \frac{sW_o(I)\left[a'+k+\left(\dfrac{a'+k}{w(I)}(1+se)\right)\right]}{\left(w(I)+1+se\right)\left(1+k+\dfrac{a'+k}{w(I)}\right)} \tag{18}$$

Therefore, the index of the constant or fixed growth of the value of production of the first sector between the two successive circuits of capital is:

$$\frac{W_1(I)}{W_0(I)} = 1 + \frac{s\left[a'+k+\left(\dfrac{a'+k}{w(I)}\right)(1+se)\right]}{\left(w(I)+1+se\right)\left(1+k+\dfrac{a'+k}{w(I)}\right)} \tag{19}$$

This series shows that the *value of production of the first sector in any circuit of capital (n) may be calculated by using the expression*:

$$W_n(I) = W_o(I)\left(1 + \frac{s\left[a' + k + \left(\frac{a'+k}{w(I)}\right)(1+se)\right]}{\left(w(I)+1+k\right)\left(1+k+\frac{a'+k}{w(I)}\right)}\right)^n \tag{20}$$

In view of the value of production of the second sector, its size in the certain circuit of capital is, according to expression (9), defined as:

$$\begin{aligned} W_1(II) = {} & W_o(II) + a'\Delta SR_1(II) + \Delta PR_1(II) \\ & + \Delta RS_1(II) + se\Delta RS_1(II) \end{aligned} \tag{21}$$

And where, according to expression (8), the growth in the value of advanced production factors is equal to the accumulation achieved in the previous circuit of capital:

$$\Delta SR_1(II) + \Delta PR_1(II) + \Delta RS_1(II) = A_o(II) \tag{22}$$

This accumulation is defined by expressions (5) and (8), so that—upon introducing expression (11)—its size is:

$$\begin{aligned} A_o(II) = {} & RS_o(I) + (1-s)V_o(I) + \Delta RS_1(I) \\ & - a'SR_o(II) - PR_o(II) + \Delta RS_1(II) \end{aligned} \tag{23}$$

The values of the individual elements on the right side of expression (23) are determined by the following expressions:

- Analogous to expression (12):

$$RS_o(I) = \frac{W_o(I)}{w(I)+1+se} \tag{24}$$

- According to expressions (16) and (17):

$$\Delta RS_1(I) = \frac{a' + k}{w(I)} \frac{sW_o(I)}{\left(w(I) + 1 + se\right)\left(1 + k + \dfrac{a' + k}{w(I)}\right)} \tag{25}$$

- Analogous to expression (12):

$$RS_o(II) = \frac{W_o(II)}{w(II) + 1 + se} \tag{26}$$

- According to expression (6):

$$PR_0(II) = kSR_0(II) \quad \text{and} \quad \Delta PR_1(II) = k\Delta SR_1(II) \tag{27}$$

- According to expression (7):

$$\left(a' + SR_o(II) + PR_0(II)\right) : RS_0(II) = w(II) \tag{28}$$

Including the expressions (26) and (27) in expression (28) results in:

$$SR_o(II) = \frac{w(II)W_o(II)}{\left(w(II) + 1 + se\right)\left(a' + k\right)} \tag{29}$$

Including expression (29) into expression (27) results in:

$$PR_o(II) = \frac{kw(II)W_o(II)}{\left(w(II) + 1 + se\right)\left(a' + k\right)} \tag{30}$$

And, according to expression (12):

$$V_o(I) = \frac{W_o(I)}{w(I) + 1 + se}$$

From this follows:

$$(1-s)V_o(I)=(1-s)\frac{W_o(I)}{w(I)+1+se} \tag{31}$$

And, according to expressions (6) and (7):

$$\Delta RS_1(II)=\Delta RS_1(II)\frac{a'+k}{w(II)} \tag{32}$$

Including the expressions (24), (31), (25), (29), (30) and (27) into expression (5) results in:

$$\frac{W_o(I)}{w(I)+1+se}\left[1+(1-s)+\frac{(a'+k)s}{w(I)\left(1+k+\dfrac{a'+k}{w(I)}\right)}\right]-\frac{w(II)W_o(II)}{w(II)+1+se} \tag{33}$$
$$=\Delta SR_1(II)(1+k)$$

Marking the entire left side of the expression (33) with Z_o, then:

$$Z_o=\Delta SR_1(II)(1+k) \tag{33a}$$

From that follows, according to expression (33a):

$$\Delta SR_1(II)=\frac{Z_o}{1+k} \tag{34}$$

Including expression (34) in expression (27) leads to:

$$\Delta PR_1(II)=\frac{kZ_o}{1+k} \tag{35}$$

While including the expression (34) in the expression (32) leads to:

$$\Delta RS_1(II) = \frac{Z_o(a'+k)}{w(II)(1+k)} \tag{36}$$

Thus the accumulation of the second sector is:

$$A_o(II) = \frac{Z_o}{1+k}\left(1+k+\frac{a'+k}{w(II)}\right) \tag{37}$$

Including the expressions (34), (35) and (36) in expression (21) results in the value of production of the second sector in the certain circuit of capital defined by the expression:

$$W_1(II) = W_o(II) + \frac{Z_o}{1+k}\left[a'+k+\frac{a'+k}{w(II)}(1+se)\right] \tag{38}$$

Thus, the *index of growth in the value of production of the second sector between the two successive circuits of capital is*:

$$\frac{W_1(II)}{W_o(II)} = 1 + \frac{\frac{Z_o}{1+k}\left[a'+k+\frac{a'+k}{w(II)}(1+se)\right]}{W_o(II)} \tag{39}$$

Generalising expressions (38) and (39) for, for instance, the *n* circuit of capital results in:

$$\frac{W_n}{W_{n-1}} = 1 + \frac{\frac{Z_{n-1}}{1+k}\left[a'+k+\frac{a'+k}{w(II)}(1+se)\right]}{W_{n-1}(II)} \tag{40}$$

Comparing expressions (19) and (40) shows that *the value of production of the first and second sector will grow at the same rate, r(I) = r(II), when*:

$$\frac{s\left[a'+k+\frac{a'+k}{w(I)}(1+se)\right]}{\left(w(I)+1+se\right)\left(1+k+\frac{a'+k}{w(I)}\right)}=\frac{\frac{Z_n-1}{1+k}\left[a'+k+\frac{a'+k}{w(II)}(1+se)\right]}{W_{n-1}} \tag{41}$$

where, according to expression (33):

$$Z_{n-1}=\frac{W_{n-1}(I)}{w(I)+1+se}\left[1+(1-s)+\frac{(a'+k)s}{w(I)\left(1+k+\frac{a'+k}{w(I)}\right)}\right]-\frac{w(II)W_{n-1}(II)}{w(II)+1+se}$$

For the purpose of simplified presentation, we mark the expressions:

$$s\left(a'+k+\frac{a'+k}{w(I)}(1+se)\right) \quad \text{with } X$$

$$\left(w(I)+1+se\right)\left(1+k+\frac{a'+k}{w(I)}\right) \quad \text{with } Y$$

$$\frac{1}{1+k}\left[a'+k+\frac{a'+k}{w(II)}(1+se)\right] \quad \text{with } M$$

$$\frac{1}{w(I)+1+se}\left[1+(1-s)+\frac{(a'+k)s}{w(I)\left(1+k+\frac{a'+k}{w(I)}\right)}\right] \quad \text{with } N$$

$$\frac{w(II)}{w(II)+1+se} \quad \text{with } L$$

Expression (41) then becomes:

$$\frac{X}{Y}=\frac{\mathrm{M}\left(NW_{n-1}(I)-LW_{n-1}(II)\right)}{W_{n-1}(II)}$$

Its solution is:

$$W_{n-1}(\mathrm{I})=\frac{X+YML}{YMN}W_{n-1}(II),(n=1,\ldots\infty) \tag{42}$$

A condition that enables the growth in the value of production of the first and second sector in the n *circuit of capital at the same rate* as in relation to their values in the (n – 1) circuit is that the relation between the value of production of the first sector in the (n – 1) circuit and the value of production of the second sector in the same circuit is the same as in:

$$W_{n-1}(I):W_{n-1}(II)=\frac{X+YML}{YMN}:1$$

This means that achieving and maintaining a growth rate of the value of production of the second sector that is equal to the growth rate of the value of production of the first sector requires that the second sector in the (n – 1) circuit and the subsequent circuits of capital forms the accumulation in the value resulting from expression (37) when comprising the expressions (33) and (42):

$$A_{n-1}(II)=\frac{\left(1+k+\dfrac{a'+k}{w(II)}\right)\left(N-L\dfrac{YMN}{X+YML}\right)}{1+k}W_{n-1}(I) \tag{43}$$

By reaching this value, the accumulation of the second sector in the value of production of the second sector becomes fixed and, following the expressions (42) and (43), is defined as:

$$\frac{A_{n-1}(II)}{W_{n-1}(II)} = \frac{1}{1+k}\left(1+k+\frac{a'+k}{w(II)}\right)\left(N\frac{X+YML}{YMN}-L\right) \quad (44)$$

This is a condition enabling the value of production of the second sector to grow in the subsequent circuits of capital at the fixed rate equal to the growth rate of the value of production of the first sector.

Consider that according to expression (13):

$$A_{n-1}(I) = s\frac{W_{n-1}(I)}{w(I)+1+se}$$

From here, it follows that *the moderate growth of the value of production in the system requires the formation of the total accumulation in the value as*:

$$A_{n-1} = \left[\frac{s}{w(I)+1+se} + \frac{\left(1+k+\frac{a'+k}{w(II)}\right)\left(N-L\frac{YMN}{X+YML}\right)}{1+k}\right]W_{n-1}(I) \quad (45)$$

All this implies that the answer to the question of whether the value of production in the first and second sector will rise at the same rate already in the second circular flow of capital reproduction or whether the rates of their growth will tend to become equal in the later (*n*th) circular flow depends essentially on the relation between the value of production of these sectors in the first circuit of capital, organic composition of these capitals, relative size of capacity, and on the rates of depreciation, accumulation and exploitation. As a consequence, the growth of production in the first and second sector will be continuously at the same rate and the growth of the total production value will be moderate only if already in the first circuit these factors that determine the value of production of the first and second sector in the circuit of capital are in a relation that enables the fulfilment of the condition defined in expression (42).

In such a situation the growth in production of the second sector may be presented as a function of the trend of production of the first sector. If we mark $W_R(I) = W_R(II)$, the value of production of the first and second sector in the circuit of capital in which they formed a relation as described in expression (43), we have:

$$W_R(II) = \frac{YMN}{X + YML} W_R(I)$$

Then, the expression Z_R may be described as:

$$Z_R = \frac{NX}{X + YML} W_R(I)$$

Therefore, including expression (33) and expression Z_R and M into expression (40) results in the following value of production of the second sector in the first circuit of capital which follows from the circuit in which the ratio of the value of production of the first and of the second sector was defined in the expression (33) and in which the production in both sectors grows at the same rate:

$$W_{R+1}(II) = \frac{YMN + NXM}{X + YML} W_R(I)$$

If this expression is included in expression (40), provided the simultaneous transformation of its denominator in accordance with expression (42), the value of production of the second sector in the circuit of capital $R + 1$ will grow in relation to the value of production in the circuit of capital R according to the following expression:

$$\frac{W_{R+1}}{W_R}(II) = 1 + \frac{X}{Y} \tag{46}$$

Therefore, the value of production of the second sector in the circuit of capital $R + n$ may be expressed as:

$$W_{R+n}(II) = \left(1 + \frac{X}{Y}\right)^{n} \frac{YMN}{X + YML} W_{R}(I) \tag{46a}$$

When the expressions X and Y are included in expression (20) it becomes:

$$W_{n}(I) = W_{o}(I)\left(1 + \frac{X}{Y}\right)^{n}$$

In this form, it expresses the value of production of the first sector in each individual circuit of capital. This means that *the total value of production of the first and second sector in the circuit of capital $R + n$ will be defined by*:

$$W_{R+n} = \left(1 + \frac{X}{Y}\right)^{n} \left(1 + \frac{YMN}{X + YML}\right) W_{R}(I) \tag{46b}$$

The condition for achieving level of production, ceteris paribus, *is to maintain the share of accumulation of the first and second sector in the value of the total production achieved in the circuit of capital R. This indicates the need that the accumulation at the end of the circuit $R + n - 1$ reaches the total level determined by expressions* (13) *and* (14):

$$A_{R+n+1} = \left(1 + \frac{X}{Y}\right)^{n-1} \left[\frac{s}{w(I) + 1 + se} + \frac{\left(1 + k + \frac{a' + k}{w(II)}\right)\left(N - L\frac{YMN}{X + YML}\right)}{1 + k} \right] W_{R}(I) \tag{47}$$

Now we can start analysing the relevant changes in the system disequilibrium, specifically the changes in:

- The share of the value of production of the first sector which at the point of transfer between the two circuits of capital cannot be realised in the system in the value of the total production of the first sector.
- The excess demand for the products of the first sector existing at the point of transfer between the two circuits of capital in relation to the value of production of the first sector in the previous circuit of capital.

Starting with the second condition of the general equilibrium of the economic system in the circumstances of expanded reproduction, the part of the value of production of the first sector at the point of transfer between the circuits of capital (n − 1) and n, which is unrealised or is realised above the value of the current production, is expressed as:

$$D_n = W_n(I) - \begin{pmatrix} ISR_n(I) + \Delta SR_{n+1}(I) + PR_n(I) + \Delta PR_{n+1}(I) \\ +ISR_n(II) + \Delta SR_{n+1}(II) + PR_n(II) + \Delta PR_{n+1}(II) \end{pmatrix},$$
$$(n = 1, \ldots \infty) \qquad (48)$$

Here, ISR_n signifies the value of the instruments of labour that were fully utilised in the nth circuit of capital and should be replaced in kind at the onset of the circuit of capital (n + 1).

Therefore, the value of ISR at the end of the nth circuit, that is at the onset of the (n + 1) circuit of capital, represents the cumulative value of the instruments of labour advanced in $(n - g\frac{1}{a'} + 1)$ circuits, provided that g represents an arithmetical progression (g = 1, 2, 3, …, G).

Therefore:

$$ISR_n = \sum_{g=1}^{G} SR_{n-g} \frac{1}{a'} + 1 \qquad (48a)$$

where:

$$n-g\frac{1}{a'}+1\geq 1$$

$G = n(a')$ and represents a whole number without decimal places.

If we include the expressions (3), (17), (27), (29), (33), (34), (35) in expression (48), it becomes:

$$D_n = W_n(I) - \left[\begin{array}{l} \sum_{g=1}^{G} \frac{W(I)W_{n-g\frac{1}{a'}+1}(I)}{\left(w(I)+1+se\right)\left(a'+k\right)} \\ +W_n(I)\left(\begin{array}{l} \frac{s(1+k)}{\left(w(I)+1+se\right)\left(1+k+\frac{a'+k}{w(I)}\right)} \\ +\frac{kw(I)}{\left(w(I)+1+se\right)\left(a'+k\right)}+N \end{array} \right) \\ +\sum_{g=1}^{G} \frac{w(II)W_{n-g\frac{1}{a'}+1}(II)}{\left(w(II)+1+se\right)\left(a'+k\right)} \\ +W_n(II)\left(\frac{kw(II)}{\left(w(II)+1+se\right)\left(a'+k\right)} \right) - L \end{array} \right] \tag{49}$$

Since according to the initial assumptions a', k, w, s and se are fixed parameters for all circuits of capital, *an absolute value D_n and its changes depend on the values and changes of the variables W_n (I) and W_n (II)*, in other words *on the rates of change and the relations among these variables.* In this case, the relation between D_n and $W_n(II)$ will change from one circuit of capital to another until the moment of establishing the equality between the rates of production growth and, thus, also a fixed relation between the value of production of the first and second sector. In line with that, by including expression (42) in expression (49), the value D_n may be expressed as a function of the value of production of the first sector only, which will enable the share of D_n in $W_n(I)$ to become fixed.

Appendix C: Conditions for Normal Reproduction in a Two-Sector Economy

Despite an extreme assumption about a system consisting of only two individual capitals, the economic categories may be normally reproduced even in conditions of a permanent disequilibrium, in relation to the conditions of equilibrium as defined by Marx. This would mean that the balance between the economic categories is established neither in the individual circuits of simple reproduction nor in the sum of the circuits of expanded reproduction.

A condition for this is the formation of the groups of relations among the relevant economic categories in the course of the capital reproduction.

Simple Reproduction

At the end of the circuit of capital where there is no replacement of the labour instruments advanced in the first and second sector $\left(1,2,\ldots,\frac{1}{a'}-1\right)$ *circuit*:

$$(1)\quad W(I) = P(PR)(I) + P(PR)(II) + D(I) \tag{50a}$$

Ž. Rohatinski, *Time and Economics*, DOI 10.1007/978-3-319-61705-3

$$(2)\quad W(II) = P_r(I) + V(I) + P_r(II) + V(II) \tag{50b}$$

$$(3)\quad P_r(I) + V(I) = N'(I) + P(PR)(II) \tag{50c}$$

where

$$D(I) = a'SR(I) + a'SR(II);$$
$$N'(I) = a'SR(II)$$

(V = surplus value).

At the end of the circuit of capital where the replacement of the labour instruments advanced in the first and second sector $\left(\frac{1}{a'}\right)$ *circuit takes place:*

(1)

$$W(I) + \sum_{1}^{1/(a'-1)} D(I) = SR(I) + P(PR)(I) + SR(II) + P(PR)(II) \tag{51a}$$

(2)

$$W(II) = P_r(I) + V(I) + P_r(II) + \mathrm{V}(\mathrm{II}) \tag{51b}$$

(3)

$$P_r(I) + \mathrm{V}(\mathrm{I}) + \mathrm{N}'(\mathrm{II}) = \mathrm{SR}(\mathrm{II}) + \mathrm{P}(\mathrm{PR})(\mathrm{II}) \tag{51c}$$

where:

$$N'(II) = \sum_{1}^{1/(a'-1)} a'SR(II)$$

Cumulatively for the circuits of capital utilising the labour instruments initially advanced in the first and second sector $\sum_{1}^{1/a'}$:

(1)

$$\sum_{1}^{1/a'} W(I) = SR(I) + \sum_{1}^{\frac{1}{a'}} P(PR)(I) + SR(II) + \sum_{1}^{1/a'} P(PR)(II) \quad (52a)$$

(2)

$$\sum_{1}^{1/a'} W(II) = \sum_{1}^{1/a'} P_r(I) + \sum_{1}^{1/a'} V(I) + \sum_{1}^{1/a'} P_r(II) + \sum_{1}^{1/a'} V(II) \quad (52b)$$

(3)

$$\sum_{1}^{1/a'} P_r(I) + V(I) + N'(II) = SR(II) + \sum_{1}^{1/a'} P(PR)(II) \quad (52c)$$

where:

$$N'(II) = \sum_{1}^{1/(a'-1)} N'(I)$$

By using the expressions:

$$SR = \frac{wW}{(w+1+se)(a'+k)}$$

$$PR = \frac{kwW}{(w+1+se)(a'+k)}$$

$$RS = \frac{W}{w+1+se}$$

$$V = \frac{seW}{w+1+se}$$

$$W = \frac{a'wW}{(w+1+se)(a'+k)} + \frac{kwW}{(w+1+se)(a'+k)} + (1+se)\frac{W}{w+1+se}$$

$$D = \frac{a'wW}{(w+1+se)(a'+k)}(I) + \frac{a'wW}{(w+1+se)(a'+k)}(II)$$

$$N'(I) = \frac{a'wW}{(w+1+se)(a'+k)}(II)$$

$$N'(II) = \sum_{1}^{1/(a'-1)} \frac{a'wW}{(w+1+se)(a'+k)}(II)$$

The relations (1–3) may be reduced to:
(1)

$$W(I) = \frac{wW(I)}{w+1+se} + \frac{wW(II)}{w+1+se} \tag{53a}$$

And if:

$$w(I) = w(II) \ i \ se(I) = se(II)$$

Then:

$$W(I) = \frac{w}{w+1+se}\left(W(I) + W(II)\right)$$

(2)

$$W(II) = \frac{(1+se)W}{w+1+se}(I) + \frac{(1+se)W}{w+1+se}(II) \tag{53b}$$

And if:

$$w(I) = w(II) \ i \ se(I) = se(II)$$

Then:

$$W(II) = \frac{1+se}{w+1+se}\left(W(I) + W(II)\right)$$

(3)

$$\frac{(1+se)W}{w+1+se}(I) = \frac{wW}{w+1+se}(II) \tag{53c}$$

And if:

$$w(I) = w(II)i \;\; se(I) = se(II)$$

Then:

$$(1+se)W(I) = wW(II)$$

Expanded Reproduction

At the end of the circuit of capital where there is no replacement of the labour instruments advanced in the first and second sector:

(1)

$$\begin{aligned} W(I) = {} & \Delta SR(I) + P(PR)(I) + \Delta P(PR)(I) + \Delta SR(II) \\ & + P(PR)(II) + \Delta P(PR)(II) + D(I) \end{aligned} \tag{54a}$$

where:

$$D(I) = a'SR(I) + a'SR(II)$$

(2)

$$\begin{aligned} W(II) = {} & P_r(I) + \Delta P_r(I) + (V-A)(I) + P_r(II) \\ & + \Delta P_r(II) + (V-A)(II) \end{aligned} \tag{54b}$$

where:

$$A(II) = \Delta SR(II) + \Delta PR(II) + \Delta RS(II)$$

(3)

$$P_r(I) + \Delta P_r(I) + (V - A)(I) = \Delta SR(II) + P(PR)(II) + \Delta P(PR)(II) + N'(I)$$

where:

$$N'(I) = a'SR(II)$$

By using the expressions:

$$A(I) = s\frac{W}{w+1+se}(I)$$

$$\Delta SR_2(I) = s\frac{W_1}{(w+1+se)\left(1+k+\frac{a'+k}{w}\right)}(I)$$

$$\Delta PR_2(I) = k\Delta SR_2(I)$$

$$\Delta RS_2(I) = \frac{a'+k}{w}\Delta SR_2(I)$$

$$A_1(II) = \Delta SR_2(II) + \Delta PR_2(II) + \Delta RS_2(II)$$

$$\Delta SR_2(II) = \frac{Z_1}{1+k}$$

$$\Delta PR_2(II) = k\Delta SR_2(II)$$

$$\Delta RS_2(II) = \frac{Z_1}{w(II)}$$

The relations (1–3) may be presented as:

(1)

$$W(I)=\frac{W}{w+1+se}\left(\frac{s(1+k)}{1+k+\frac{a'+k}{w}}+\frac{kw}{a'+k}\right)(I)+Z+\frac{kwW}{(w+1+se)(a'+k)}(II)+D(I) \tag{55a}$$

(2)

$$W(II)=\frac{W}{w+1+se}\left(1+se+\frac{s(a'+k)}{w\left(1+k+\frac{a'+k}{w}\right)}-s\right)(I)+\frac{(1+se)W}{w+1+se}(II)-Z \tag{55b}$$

(3)

$$\frac{W}{w+1+se}\left(1+se+\frac{s(a'+k)}{w\left(1+k+\frac{a'+k}{w}\right)}-s\right)(I)=Z+\frac{kwW}{(w+1+se)(a'+k)}(II)+N'(I) \tag{55c}$$

At the end of the circuit of capital where the replacement of labour instruments advanced in the first circuit of capital of the first and second sector $\left(\frac{1}{a'}\right)$ *circuit is carried out:*

(1)

$$W_{\frac{1}{a'}}(I) = \sum_{1}^{1/(a'-1)} D(I) = SR_1(I) + SR_1(II) + \Delta SR_{\frac{1}{a'+1}}(I) + \Delta SR_{\frac{1}{a'+1}}(II) + P(PR)_{\frac{1}{a'}}(I) + P(PR)_{\frac{1}{a'}}(II) + \Delta P(PR)_{\frac{1}{a'+1}}(I) + \Delta P(PR)_{\frac{1}{a'+1}}(II) + AM(I) + AM(II) \quad (56a)$$

where, $D(I)$ according to expression (49):

$$AM(I) + AM(II) = AM = a'\left(\Delta SR_{\frac{1}{a'}}(I) + \Delta SR_{\frac{1}{a'}}(II)\right) + 2a'\left(\Delta SR_{\frac{1}{a'-1}}(I) + \Delta SR_{\frac{1}{a'-1}}(II)\right) + \ldots + na'\left(\Delta SR_{1/(a'-n)}(I) + \Delta SR_{1/(a'-n)}(II)\right)$$

AM = value of labour instruments utilised in the circuits of capital $1/(a'-n)$ to $1/a'$ not replaced at the end of the $1/a'$ circuit of capital (when $n = \frac{1}{a'} - 1$).

(2)

$$W_{\frac{1}{a'}}(II) = P_{r_{\frac{1}{a'}}}(I) + \Delta P_{r_{\frac{1}{a'+1}}}(I) + (V-A)_{\frac{1}{a'}}(I) + P_{r_{\frac{1}{a'}}}(II) + \Delta P_{r_{\frac{1}{a'+1}}}(II) + (V-A)_{\frac{1}{a'}}(II) \quad (56b)$$

(3)

$$P_{r_{\frac{1}{a'}}(I)} + \Delta P_{r_{\frac{1}{a'+1}}}(I) + (V-A)_{\frac{1}{a'}}(I) + N'_{\frac{1}{a'}}(II) = \Delta SR_1(II) + \Delta SR_{\frac{1}{a'+1}}(II) + P(PR)_{\frac{1}{a'}}(II) + \Delta P(PR)_{\frac{1}{a'+1}}(II) \quad (56c)$$

where:

$$N'_{\frac{1}{a'}}(II) = SR_{\frac{1}{a'-n}}(II) - \begin{pmatrix} a'SR_{\frac{1}{a'-n}} + a'SR_{\frac{1}{a'-n+1}} \\ +a'SR_{\frac{1}{a'-n+2}} + \ldots + a'SR_{\frac{1}{a'}}(II) \end{pmatrix}$$

Cumulatively for the circuits of capital utilising the labour instruments initially advanced in the first and second sector $\sum_{1}^{1/a'}$ *:*

(1)

$$\sum_{i=1}^{1/a'} W_i(I) = SR_1 + \sum_{i=2}^{1/(a'+1)} \Delta SR_i(I+II) + \sum_{i=2}^{1/(a'+1)} \Delta P(PR)_i(I+II) + \sum_{1(a'-n)}^{1/a'} D \tag{57a}$$

where:

$$\sum_{1(a'-n)}^{1/a'} D = \left(a'SR_{1/a'} + 2a'SR_{1/(a'-n+1)} + \ldots + na'SR_{\frac{1}{a'-n}} \right)(I+II)$$

(2)

$$\sum_{i=1}^{1/a'} W_i(II) = \sum_{i=2}^{1/(a'+1)} P_{ri}(I+II) + \sum_{1}^{1/a'}(V-A)(I+II) \tag{57b}$$

(3)

$$\sum_{i=2}^{1/(a'+1)} P_{ri}(I) + \sum_{1}^{1/a'}(V-A)(I) = SR_1(II) + \sum_{i=2}^{1/(a'+1)} \Delta SR_1(II) + \sum_{i=2}^{1/(a'+1)} P(PR)(II) + S_O N'(I) \tag{57c}$$

where:

$$S_O N'(I) = a'SR_{\frac{1}{a'}}(II) + 2a'SR_{\frac{1}{a'-1}}(II)$$

Appendix D: Conditions for Equilibrium of a Two-Sector Economy in Functional Time

Determining the modalities of establishing the equilibrium in a two-sector system requires expressing all of the time attributes of the capital reproduction in the first and second sector as functions of the duration of labour periods in these sectors. Therefore:

- Duration of labour period in the first sector, $n(I)$.
- Duration of labour period in the second sector, $n(II)$ or $vn(I)$, where $v < 1$.
- Period for which the wages are paid in the first sector is $x(I)$ times longer than the duration of a labour period in the first sector, $x(I)n(I)$.
- Analogously, the duration of the period for which the wages are paid in the second sector, $x(II)n(II)$.
- Duration of turnover of the total capital advanced in the first sector is $y(I)$ times longer than the duration of a single labour period in the first sector, $y(I)n(I)$.
- Analogously, the duration of turnover of the total capital advanced in the second sector, $y(II)n(II)$.

On this basis, *the value of production of the first sector in a single labour period of reproduction of the capital advanced in this sector* may be expressed as:

Ž. Rohatinski, *Time and Economics*, DOI 10.1007/978-3-319-61705-3

$$W(I) = n(I)PR(I) + (1+se)\frac{1}{x(I)}n(I)RS(I)$$
$$+\frac{1}{y(I)}n(I)SR(I)$$

And *the value of production of the second sector in a single labour period of this sector* may be expressed as:

$$W(II) = n(II)PR(II) + (1+se)\frac{1}{x(II)}n(II)RS(II)$$
$$+\frac{1}{y(II)}n(II)SR(II)$$

Analogous to earlier deliberations, the relation between the value of the production factors *PR*, *RS* and *SR* is defined by the parameters$k$, w, i, a' whose analytical derivatives in this case read as follows:

- For the first sector:

$$w(I)n(I) = \frac{x(I)n(I)\left(k(I)n(I) + \frac{1}{y(I)}n(I)\right)SR(I)}{RS(I)}$$

- For the second sector:

$$w(II)n(II) = \frac{x(II)n(II)\left(k(II)n(II) + \frac{1}{y(II)}n(II)\right)SR(II)}{RS(II)}$$

From this follows that *the values of production of the first and the second sectorin a single labour period of the first and the second sector* are defined by the following expressions:

$$W(I)=\left[k(I)n(I)+(1+se)\frac{k(I)n(I)+\frac{1}{y(I)}n(I)}{w(I)n(I)}+\frac{1}{y(I)}n(I)\right]SR(I)$$

$$W(II)=\left[\begin{array}{l}k(II)n(II)+(1+se)\frac{k(II)n(II)+\frac{1}{y(II)}n(II)}{w(II)n(II)}\\+\frac{1}{y(II)}n(II)\end{array}\right]SR(II)$$

The *normal exchange between sectors in such circumstances requires that in the period y (II) n (II) the following condition of equilibrium be met*:

$$\begin{aligned}&\left(y(II)n(II)k(II)n(II)+1\right)SR(II)\\&=\frac{y(II)n(II)(1+se)\left[k(I)n(I)+\frac{1}{y(I)}n(I)\right]}{x(I)n(I)w(I)n(I)}SR(I)\end{aligned}$$

From here, *the required equilibrium relation between the value of the instruments of labour employed in the first and second sector* follows as:

$$SR(II)=\frac{y(II)n(II)(1+se)\left[k(I)n(I)+\frac{1}{y(I)}n(I)\right]}{x(I)n(I)w(I)n(I)\left(y(II)n(II)k(II)n(II)+1\right)}SR(I)$$

Upon including this expression of the relation between the durations of labour periods in the first and second sector $n(I)=\frac{1}{V}n(II)$, *establishing the equilibrium in the system requires the formation of the following relation between the value of the labour instruments to be employed in the first and second sector of capital reproduction*:

$$SR(II) = \frac{y(II)n(II)(1+se)\left[k(I)\frac{1}{v}n(I) + \frac{1}{y(I)}\frac{1}{v}n(I)\right]}{x(I)\frac{1}{v}n(II)w(I)\frac{1}{v}n(II)\left[y(II)n(II)k(II)n(II)+1\right]}SR(I)$$

Or, simplified:

$$SR(II) = CSR(I)$$

Where C represents the equilibrium coefficient of the value of the labour instruments to be employed in the second sector in relation to the unit of the value of the labour instruments employed in the first sector:

$$C = \frac{y(II)n(II)(1+se)\left[k(I)\frac{1}{v}n(I) + \frac{1}{y(I)}\frac{1}{v}n(II)\right]}{x(I)\frac{1}{v}n(II)w(I)\frac{1}{v}n(II)\left[y(II)n(II)k(II)n(II)+1\right]}$$

Accordingly, *the total advanced money capital N should be distributed to the capital advanced in the first sector N(I) and the capital advanced in the second sector N(II)* and within these the capital advanced for employment of individual production factors in accordance with the following expressions:

$$N(I) = \left[1 + k(I)n(I) + \frac{x(I)n(I)\left(k(I)n(I) + \frac{1}{y(I)}n(I)\right)}{w(I)n(I)}\right]SR(I)$$

$$N(II)=\left[\begin{array}{l}1+k(II)\frac{1}{v}n(II)\\ +\frac{x(II)n(II)\left(k(II)n(II)+\frac{1}{y(II)}n(II)\right)}{w(II)n(II)}\end{array}\right]CSR(I)$$

The value of the objects of labour to be advanced to the second sector at the beginning of the first period of a single circular flow of reproduction of capital in this sector is defined as:

$$PR(II)=k(II)\frac{1}{v}n(II)CSR(I)$$

And *not* by the expression:

$$PR(II)=k(II)n(II)CSR(I)$$

This second expression is erroneous because the duration of a labour period in production of the means of production the first sector is $\frac{1}{V}$ times longer than the duration of a labour period in the second sector. For this reason, the second sector must advance the means of production in the size $\frac{1}{V}$ times larger than those effectively utilised in a single labour period of the second sector, since there will be no chance of acquiring them again until the end of the labour period in the first sector.

If we, for simplification, mark the expression:

$$1+k(I)+n(I)+\frac{x(I)n(I)\left[k(I)n(I)+\frac{1}{y(I)}n(I)\right]}{w(I)n(I)}\text{ with }A$$

And the expression:

$$1+k(II)+\frac{1}{v}n(II)+\frac{x(II)n(II)\left[k(II)n(II)+\frac{1}{y(II)}n(II)\right]}{w(II)n(II)} \quad \text{with } B$$

Then:

$$N(I)=ASR(I)$$
$$N(II)=BCSR(I)$$

And from here follows:

$$N=(A+BC)SR(I)$$

This enables us to determine the equilibrium distribution of the total advanced money capital to the individual production factors in the first and second sector, and the equilibrium value of production in each individual sector. Starting from the analytical derivatives for determining the values of *N* (*I*) and *N* (*II*), the result is:

$$PR(I)=k(I)n(I)SR(I)$$

$$RS(I)=\frac{x(I)n(I)\left(k(I)n(I)+\frac{1}{y(I)}n(I)\right)}{w(I)n(I)}SR(I)$$

$$SR(II)=CSR(I)$$

$$PR(II)=k(II)\frac{1}{v}n(II)CSR(I) \quad \textit{provided that}: \ P(PR)=vPR$$

$$RS(II)=\frac{x(II)n(II)\left(k(II)n(II)+\frac{1}{y(II)}n(II)\right)}{w(II)n(II)}CSR(I)$$

Considering that $SR(I)=\frac{1}{A+BC}N$, *the values of production of the first and second sector in a single labour period of reproduction in these sectors required for establishing the cumulative equilibrium in a system* are defined with the following expressions:

$$W(I)=\left[\begin{array}{l} k(I)n(I)+(1+se)\dfrac{k(I)n(I)+\dfrac{1}{y(I)}n(I)}{w(I)n(I)} \\ +\dfrac{1}{y(I)}n(I) \end{array}\right]\frac{1}{A+BC}N$$

$$W(II)=\left[\begin{array}{l} k(II)n(II)+(1+se)\dfrac{k(II)n(II)+\dfrac{1}{y(II)}n(II)}{w(II)n(II)} \\ +\dfrac{1}{y(II)}n(II) \end{array}\right]\frac{C}{A+BC}N$$

In that way, all elements of the flows of the capital reproduction in the first and second sector are functionally determined by the size of the total money capital advanced in the system, technical and technological characteristics of the employed factors of production, specific functional time attributes of the course of reproduction in the first, that is, in the second sector, and by the relation between the durations of the labour periods as time measurement units in these sectors.

Now we can determine *the period in which the cumulative equilibrium in the system is established*. This equilibrium is a specific constellation of the causal and structural relations between economic categories, enabling the normal reproduction in the system, while all elements of capital reproduction in the first and the second sector are functionally determined by the specific characteristics of the time flows of the reproduction expressed in the units of duration of the labour periods in the first and second sector. *The cumulative equilibrium in a system may be established only in the period within which the times of the turnover of the total capital advanced in the first and second sector expressed in the units of duration of the labour periods in these sectors fully overlap*. Therefore, the duration of this period is defined as:

$$My(I)n(I) = Ny(II)n(II)$$

Considering $n(I) = \frac{1}{v}n(II)$, the duration of the labour period in the first second is:

$$My(I)n(I) = Nvy(II)n(I)$$

And in the second sector: labour period

$$M\frac{1}{v}y(I)n(II) = Ny(II)n(II)$$

Here the expressions of M and N are whole numbers greater than zero, used to establish the equality between the values of $y(I)n(I)$ and $y(II)n(II)$. They determine the number of turnovers of the total capital advanced in the first (M) and the second sector (N) which must be realised to equalise the sum of the turnover times of these capitals, and thus establish the value-time equilibrium in a system. Considering that in the rule $y(I)n(I) \neq y(II)n(II)$ and $v \neq 1$, *the cumulative equilibrium in the system may be established for a period the duration of which, expressed in the units of labour periods of the first or the second sector, is determined as the lowest common multiple of the numbers* $y(I)n(I)$, $y(II)n(II)$ *and* v. Therefore:

$$M = \frac{y(I)n(I)y(II)n(II)v}{y(I)n(I)} = vy(II)n(II)$$

$$N = \frac{y(I)n(I)y(II)n(II)v}{y(II)n(II)} = y(I)n(I)$$

From here, it follows that *the cumulative conditions of the system equilibrium are defined by the following expressions*:

(1)

$$\sum_{1}^{MN} P_r(I) + \sum_{1}^{MN} V(I) = \sum_{1}^{MN\frac{1}{v}} P(PR)(II) + \sum_{1}^{N} SR(II)$$

(2)

$$\sum_{1}^{M} SR(I) + \sum_{1}^{MN} P(PR)(I) + \sum_{1}^{N} SR(II) + \sum_{1}^{MN\frac{1}{v}} P(PR)(II) = \sum_{1}^{MN} W(I)$$

(3)

$$\sum_{1}^{MN} P_r(I) + \sum_{1}^{MN} V(I) + \sum_{1}^{MN\frac{1}{v}} P_r(II) + \sum_{1}^{MN\frac{1}{v}} V(II) = \sum_{1}^{MN\frac{1}{v}} W(II)$$

Index[1]

[1]Note: Page numbers followed by "n" refers to note.

Ž. Rohatinski, *Time and Economics*, DOI 10.1007/978-3-319-61705-3

The manufacturer's authorised representative in the EU is Springer Nature Customer Service Centre GmbH, Europaplatz 3, 69115 Heidelberg, Germany. If you have any concerns regarding our products, please contact ProductSafety@springernature.com

Printed and bound by CPI Group (UK) Ltd, Croydon, CR0 4YY
01/10/2026
02225967-0010